BIG IDEAS
MATH®
Modeling Real Life

Grade K

Volume 2

Ron Larson
Laurie Boswell

BIG IDEAS
LEARNING®

Erie, Pennsylvania
BigIdeasLearning.com

Big Ideas Learning, LLC
1762 Norcross Road
Erie, PA 16510-3838
USA

For product information and customer support, contact Big Ideas Learning
at 1-877-552-7766 or visit us at BigIdeasLearning.com.

Cover Image
Paul Lampard /123RF.com, enmyo/Shutterstock.com

Big Ideas Learning and Big Ideas Math are registered trademarks of Larson Texts, Inc.

Printed in the U.S.A.

ISBN 13: 978-1-63598-872-7

7 8 9 10—23

About the Authors

Ron Larson

Ron Larson, Ph.D., is well known as the lead author of a comprehensive program for mathematics that spans school mathematics and college courses. He holds the distinction of Professor Emeritus from Penn State Erie, The Behrend College, where he taught for nearly 40 years. He received his Ph.D. in mathematics from the University of Colorado. Dr. Larson's numerous professional activities keep him actively involved in the mathematics education community and allow him to fully understand the needs of students, teachers, supervisors, and administrators.

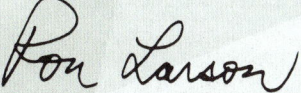

Laurie Boswell

Laurie Boswell, Ed.D., is the former Head of School at Riverside School in Lyndonville, Vermont. In addition to textbook authoring, she provides mathematics consulting and embedded coaching sessions. Dr. Boswell received her Ed.D. from the University of Vermont in 2010. She is a recipient of the Presidential Award for Excellence in Mathematics Teaching and is a Tandy Technology Scholar. Laurie has taught math to students at all levels, elementary through college. In addition, Laurie has served on the NCTM Board of Directors and as a Regional Director for NCSM. Along with Ron, Laurie has co-authored numerous math programs and has become a popular national speaker.

Dr. Ron Larson and Dr. Laurie Boswell began writing together in 1992. Since that time, they have authored over four dozen textbooks. This successful collaboration allows for one voice from Kindergarten through Algebra 2.

Contributors, Reviewers, and Research

Big Ideas Learning would like to express our gratitude to the mathematics education and instruction experts who served as our advisory panel, contributing specialists, and reviewers during the writing of *Big Ideas Math: Modeling Real Life*. Their input was an invaluable asset during the development of this program.

Contributing Specialists and Reviewers

- **Sophie Murphy**, Ph.D. Candidate, Melbourne School of Education, Melbourne, Australia
  Learning Targets and Success Criteria Specialist and Visible Learning Reviewer

- **Linda Hall**, Mathematics Educational Consultant, Edmond, OK
  Advisory Panel

- **Michael McDowell**, Ed.D., Superintendent, Ross, CA
  Project-Based Learning Specialist

- **Kelly Byrne**, Math Supervisor and Coordinator of Data Analysis, Downingtown, PA
  Advisory Panel

- **Jean Carwin**, Math Specialist/TOSA, Snohomish, WA
  Advisory Panel

- **Nancy Siddens**, Independent Language Teaching Consultant, Las Cruces, NM
  English Language Learner Specialist

- **Kristen Karbon**, Curriculum and Assessment Coordinator, Troy, MI
  Advisory Panel

- **Kery Obradovich**, K–8 Math/Science Coordinator, Northbrook, IL
  Advisory Panel

- **Jennifer Rollins**, Math Curriculum Content Specialist, Golden, CO
  Advisory Panel

- **Becky Walker**, Ph.D., School Improvement Services Director, Green Bay, WI
  Advisory Panel and Content Reviewer

- **Deborah Donovan**, Mathematics Consultant, Lexington, SC
  Content Reviewer

- **Tom Muchlinski**, Ph.D., Mathematics Consultant, Plymouth, MN
  Content Reviewer and Teaching Edition Contributor

- **Mary Goetz**, Elementary School Teacher, Troy, MI
  Content Reviewer

- **Nanci N. Smith**, Ph.D., International Curriculum and Instruction Consultant, Peoria, AZ
  Teaching Edition Contributor

- **Robyn Seifert-Decker**, Mathematics Consultant, Grand Haven, MI
  Teaching Edition Contributor

- **Bonnie Spence**, Mathematics Education Specialist, Missoula, MT
  Teaching Edition Contributor

- **Suzy Gagnon**, Adjunct Instructor, University of New Hampshire, Portsmouth, NH
  Teaching Edition Contributor

- **Art Johnson**, Ed.D., Professor of Mathematics Education, Warwick, RI
  Teaching Edition Contributor

- **Anthony Smith**, Ph.D., Associate Professor, Associate Dean, University of Washington Bothell, Seattle, WA
  Reading and Writing Reviewer

- **Brianna Raygor**, Music Teacher, Fridley, MN
  Music Reviewer

- **Nicole Dimich Vagle**, Educator, Author, and Consultant, Hopkins, MN
  Assessment Reviewer

- **Janet Graham**, District Math Specialist, Manassas, VA
  Response to Intervention and Differentiated Instruction Reviewer

- **Sharon Huber**, Director of Elementary Mathematics, Chesapeake, VA
  Universal Design for Learning Reviewer

Student Reviewers

- T.J. Morin
- Alayna Morin
- Ethan Bauer
- Emery Bauer
- Emma Gaeta
- Ryan Gaeta
- Benjamin SanFrotello
- Bailey SanFrotello
- Samantha Grygier
- Robert Grygier IV
- Jacob Grygier
- Jessica Urso
- Ike Patton
- Jake Lobaugh
- Adam Fried
- Caroline Naser
- Charlotte Naser

Research

Ron Larson and Laurie Boswell used the latest in educational research, along with the body of knowledge collected from expert mathematics instructors, to develop the *Modeling Real Life* series. The pedagogical approach used in this program follows the best practices outlined in the most prominent and widely accepted educational research, including:

- *Visible Learning*
  John Hattie © 2009

- *Visible Learning for Teachers*
  John Hattie © 2012

- *Visible Learning for Mathematics*
  John Hattie © 2017

- *Principles to Actions: Ensuring Mathematical Success for All*
  NCTM © 2014

- *Adding It Up: Helping Children Learn Mathematics*
  National Research Council © 2001

- *Mathematical Mindsets: Unleashing Students' Potential through Creative Math, Inspiring Messages and Innovative Teaching*
  Jo Boaler © 2015

- *What Works in Schools: Translating Research into Action*
  Robert Marzano © 2003

- *Classroom Instruction That Works: Research-Based Strategies for Increasing Student Achievement*
  Marzano, Pickering, and Pollock © 2001

- *Principles and Standards for School Mathematics*
  NCTM © 2000

- *Rigorous PBL by Design: Three Shifts for Developing Confident and Competent Learners*
  Michael McDowell © 2017

- *Universal Design for Learning Guidelines*
  CAST © 2011

- *Rigor/Relevance Framework®*
  International Center for Leadership in Education

- *Understanding by Design*
  Grant Wiggins and Jay McTighe © 2005

- Achieve, ACT, and The College Board

- *Elementary and Middle School Mathematics: Teaching Developmentally*
  John A. Van de Walle and Karen S. Karp © 2015

- *Evaluating the Quality of Learning: The SOLO Taxonomy*
  John B. Biggs & Kevin F. Collis © 1982

- *Unlocking Formative Assessment: Practical Strategies for Enhancing Students' Learning in the Primary and Intermediate Classroom*
  Shirley Clarke, Helen Timperley, and John Hattie © 2004

- *Formative Assessment in the Secondary Classroom*
  Shirley Clarke © 2005

- *Improving Student Achievement: A Practical Guide to Assessment for Learning*
  Toni Glasson © 2009

Mathematical Processes and Proficiencies

Big Ideas Math: Modeling Real Life reinforces the Process Standards from NCTM and the Five Strands of Mathematical Proficiency endorsed by the National Research Council. With *Big Ideas Math*, students get the practice they need to become well-rounded, mathematically proficient learners.

Problem Solving/Strategic Competence

- *Think & Grow: Modeling Real Life* examples use problem-solving strategies, such as drawing a picture, circling knowns, and underlining unknowns.
- Real-life problems are provided to help students learn to apply the mathematics that they are learning to everyday life.
- Real-life problems help students use the structure of mathematics to break down and solve more difficult problems.

Reasoning and Proof/Adaptive Reasoning

- *Explore & Grows* allow students to investigate math and make conjectures.
- Questions ask students to explain their reasoning.

Communication

- Cooperative learning opportunities support precise communication.
- *Apply and Grow: Practice* exercises allow students to demonstrate their understanding of the lesson up to that point.
- *ELL Support* notes provide insights into how to support English learners.

Connections

- Prior knowledge is continually brought back and tied in with current learning.
- Performance Tasks tie the topics of a chapter together into one extended task.
- Real-life problems incorporate other disciplines to help students see that math is used across content areas.

Representations/Productive Disposition

- Real-life problems are translated into pictures, diagrams, tables, equations, or graphs to help students analyze relations and to draw conclusions.
- Visual problem-solving models help students create a coherent representation of the problem.
- Multiple representations are presented to help students move from concrete to representative and into abstract thinking.
- *Learning Targets* and *Success Criteria* at the start of each chapter and lesson help students understand what they are going to learn.
- Real-life problems incorporate other disciplines to help students see that math is used across content areas.

Conceptual Understanding

- *Explore & Grows* allow students to investigate math to understand the reasoning behind the rules.

Procedural Fluency

- Skill exercises are provided to continually practice fundamental skills.
- Prior knowledge is continually brought back and tied in with current learning.

Meeting Proficiency and Major Topics

Meeting Proficiency

As standards shift to prepare students for college and careers, the importance of focus, coherence, and rigor continues to grow.

FOCUS — *Big Ideas Math: Modeling Real Life* emphasizes a narrower and deeper curriculum, ensuring students spend their time on the major topics of each grade.

COHERENCE — The program was developed around coherent progressions from Kindergarten through eighth grade, guaranteeing students develop and progress their foundational skills through the grades while maintaining a strong focus on the major topics.

RIGOR — *Big Ideas Math: Modeling Real Life* uses a balance of procedural fluency, conceptual understanding, and real-life applications. Students develop conceptual understanding in every *Explore and Grow*, continue that development through the lesson while gaining procedural fluency during the *Think and Grow*, and then tie it all together with *Think and Grow: Modeling Real Life*. Every set of practice problems reflects this balance, giving students the rigorous practice they need to be college- and career-ready.

Major Topics in Kindergarten

Counting and Cardinality
- Know number names and the count sequence.
- Count to tell the number of objects.
- Compare numbers.

Operations and Algebraic Thinking
- Understand addition as putting together and adding to, and understand subtraction as taking apart and taking from.

Number and Operations in Base Ten
- Work with numbers 11–19 to gain foundations for place value.

Use the color-coded Table of Contents to determine where the major topics, supporting topics, and additional topics occur throughout the curriculum.

- 🟩 Major Topic
- 🟦 Supporting Topic
- 🟨 Additional Topic

① Count and Write Numbers 0 to 5

Vocabulary 2

- 1.1 Model and Count 1 and 2 3
- 1.2 Understand and Write 1 and 2 9
- 1.3 Model and Count 3 and 4 15
- 1.4 Understand and Write 3 and 4 21
- 1.5 Model and Count 5 27
- 1.6 Understand and Write 5 33
- 1.7 The Concept of Zero 39
- 1.8 Count and Order Numbers to 5 45

Performance Task: Farm Animals 51

Game: Number Land 52

Chapter Practice 53

② Compare Numbers 0 to 5

Vocabulary 58

- 2.1 Equal Groups 59
- 2.2 Greater Than 65
- 2.3 Less Than 71
- 2.4 Compare Groups to 5 by Counting 77
- 2.5 Compare Numbers to 5 83

Performance Task: Games 89

Game: Toss and Compare 90

Chapter Practice 91

- ■ Major Topic
- ■ Supporting Topic
- ■ Additional Topic

3 Count and Write Numbers 6 to 10

Vocabulary 96

■ 3.1 Model and Count 6 97

■ 3.2 Understand and Write 6 103

■ 3.3 Model and Count 7 109

■ 3.4 Understand and Write 7 115

■ 3.5 Model and Count 8 121

■ 3.6 Understand and Write 8 127

■ 3.7 Model and Count 9 133

■ 3.8 Understand and Write 9 139

■ 3.9 Model and Count 10 145

■ 3.10 Understand and Write 10 151

■ 3.11 Count and Order Numbers to 10 157

Performance Task: Safari Animals 163

Game: Number Land 164

Chapter Practice 165

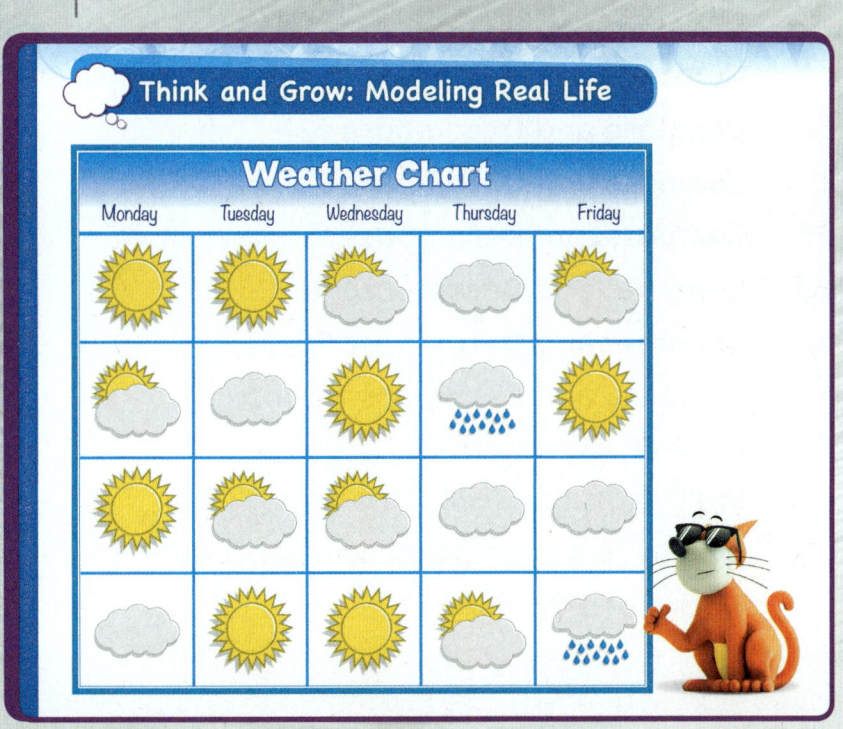

Compare Numbers to 10

Vocabulary 170

■ **4.1** Compare Groups to 10 by Matching 171

■ **4.2** Compare Groups to 10 by Counting 177

■ **4.3** Compare Numbers to 10 183

■ **4.4** Classify Objects into Categories 189

■ **4.5** Classify and Compare by Counting 195

Performance Task: Toys 201

Game: Toss and Compare 202

Chapter Practice 203

Cumulative Practice 207

Compose and Decompose Numbers to 10

Vocabulary 212

■ **5.1** Partner Numbers to 5 213

■ **5.2** Use Number Bonds to Represent
Numbers to 5 219

■ **5.3** Compose and Decompose 6 225

■ **5.4** Compose and Decompose 7 231

■ **5.5** Compose and Decompose 8 237

■ **5.6** Compose and Decompose 9 243

■ **5.7** Compose and Decompose 10 249

■ **5.8** Compose and Decompose Using
a Group of 5 255

Performance Task: Insects 261

Game: Number Bond Spin and Cover 262

Chapter Practice 263

■ Major Topic
■ Supporting Topic
■ Additional Topic

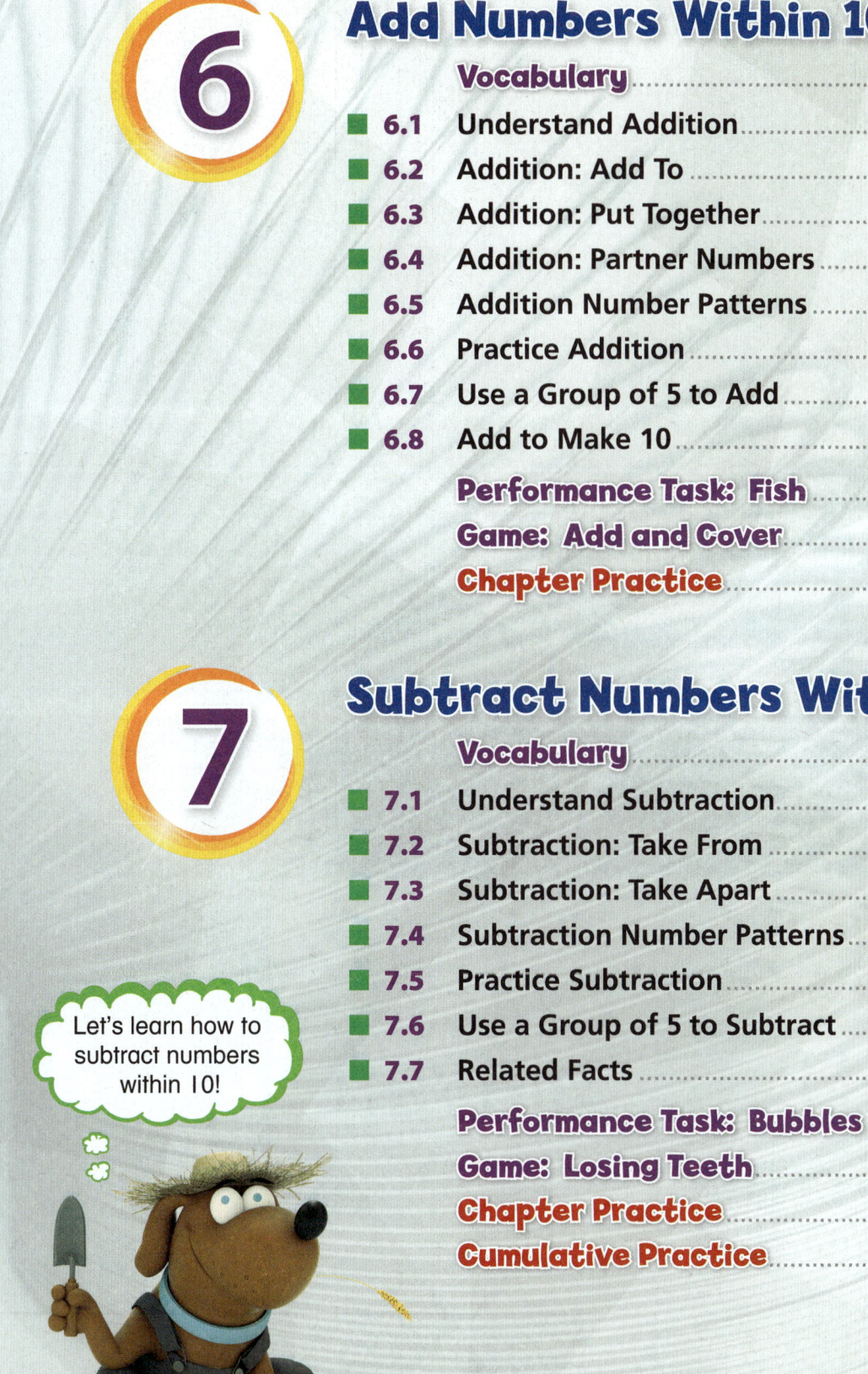

6 Add Numbers Within 10

Vocabulary 268
◾ 6.1 Understand Addition 269
◾ 6.2 Addition: Add To 275
◾ 6.3 Addition: Put Together 281
◾ 6.4 Addition: Partner Numbers 287
◾ 6.5 Addition Number Patterns 293
◾ 6.6 Practice Addition 299
◾ 6.7 Use a Group of 5 to Add 305
◾ 6.8 Add to Make 10 311

Performance Task: Fish 317
Game: Add and Cover 318
Chapter Practice 319

7 Subtract Numbers Within 10

Vocabulary 324
◾ 7.1 Understand Subtraction 325
◾ 7.2 Subtraction: Take From 331
◾ 7.3 Subtraction: Take Apart 337
◾ 7.4 Subtraction Number Patterns 343
◾ 7.5 Practice Subtraction 349
◾ 7.6 Use a Group of 5 to Subtract 355
◾ 7.7 Related Facts 361

Performance Task: Bubbles 367
Game: Losing Teeth 368
Chapter Practice 369
Cumulative Practice 373

Let's learn how to subtract numbers within 10!

8

Represent Numbers 11 to 19

Vocabulary 378
■ **8.1** Identify Groups of 10 379
■ **8.2** Count and Write 11 and 12 385
■ **8.3** Understand 11 and 12 391
■ **8.4** Count and Write 13 and 14 397
■ **8.5** Understand 13 and 14 403
■ **8.6** Count and Write 15 409
■ **8.7** Understand 15 415
■ **8.8** Count and Write 16 and 17 421
■ **8.9** Understand 16 and 17 427
■ **8.10** Count and Write 18 and 19 433
■ **8.11** Understand 18 and 19 439

Performance Task: Stars 445
Game: Number Flip and Find 446
Chapter Practice 447

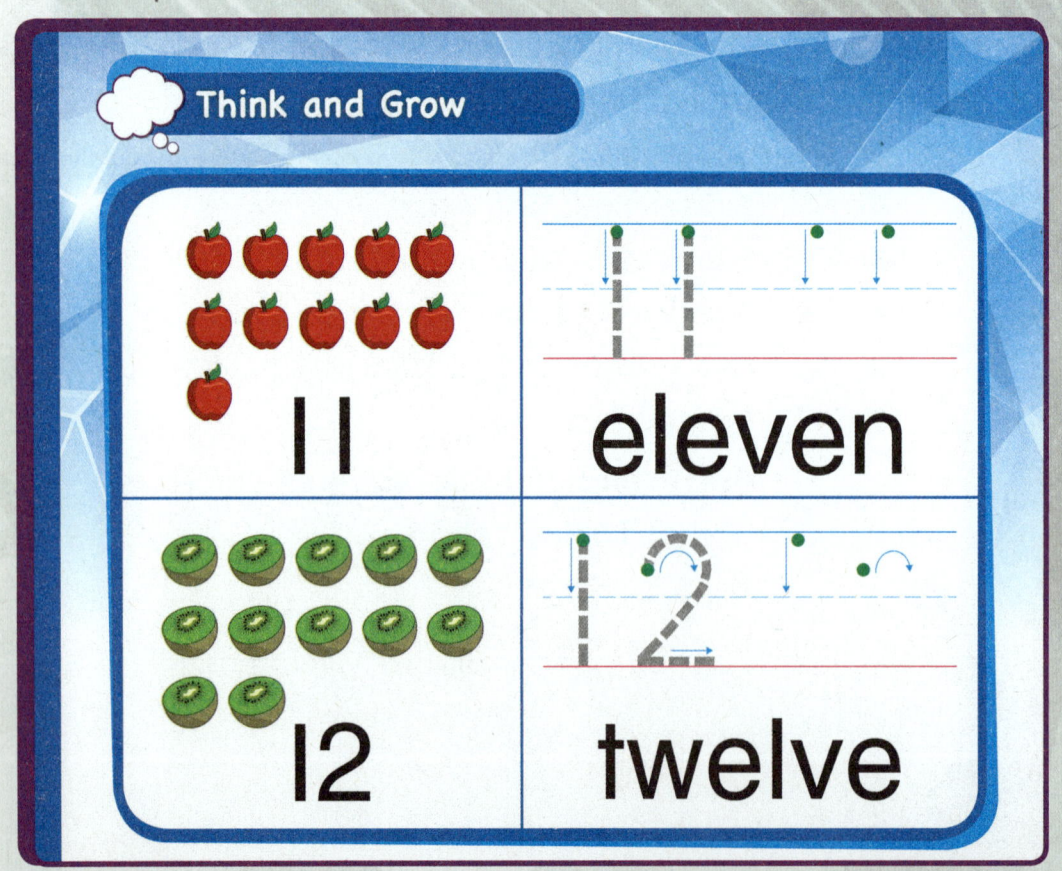

9 Count and Compare Numbers to 20

Vocabulary 454

■ **9.1** Model and Count 20 455

■ **9.2** Count and Write 20 461

■ **9.3** Count to Find How Many 467

■ **9.4** Count Forward from Any Number to 20 473

■ **9.5** Order Numbers to 20 479

■ **9.6** Compare Numbers to 20 485

Performance Task: Fruit 491

Game: Number Boss 492

Chapter Practice 493

10 Count to 100

Vocabulary 498

■ **10.1** Count to 30 by Ones 499

■ **10.2** Count to 50 by Ones 505

■ **10.3** Count to 100 by Ones 511

■ **10.4** Count to 100 by Tens 517

■ **10.5** Count by Tens and Ones 523

■ **10.6** Count by Tens from a Number 529

Performance Task: Party Supplies 535

Game: Hundred Chart Puzzle 536

Chapter Practice 537

Cumulative Practice 541

Identify Two-Dimensional Shapes

Vocabulary 546

■ **11.1** Describe Two-Dimensional Shapes 547

■ **11.2** Triangles 553

■ **11.3** Rectangles 559

■ **11.4** Squares 565

■ **11.5** Hexagons and Circles 571

■ **11.6** Join Two-Dimensional Shapes 577

■ **11.7** Build Two-Dimensional Shapes 583

Performance Task: Pets 589

Game: Shape Flip and Find 590

Chapter Practice 591

Identify Three-Dimensional Shapes and Positions

Vocabulary 596

■ **12.1** Two- and Three-Dimensional Shapes 597

■ **12.2** Describe Three-Dimensional Shapes 603

■ **12.3** Cubes and Spheres 609

■ **12.4** Cones and Cylinders 615

■ **12.5** Build Three-Dimensional Shapes 621

■ **12.6** Positions of Solid Shapes 627

Performance Task: Recycling 633

Game: Solid Shapes: Spin and Cover 634

Chapter Practice 635

■ Major Topic
■ Supporting Topic
■ Additional Topic

13 Measure and Compare Objects

Vocabulary 640

■ 13.1 Compare Heights 641

■ 13.2 Compare Lengths 647

■ 13.3 Use Numbers to Compare Lengths 653

■ 13.4 Compare Weights 659

■ 13.5 Use Numbers to Compare Weights 665

■ 13.6 Compare Capacities 671

■ 13.7 Describe Objects by Attributes 677

Performance Task: Rainwater 683

Game: Measurement Boss 684

Chapter Practice 685

Cumulative Practice 689

Glossary A1

Index A15

8 Represent Numbers 11 to 19

- What kinds of objects can you see in the night sky?

- How many bright stars are in each group shown? How many are there in all?

Chapter Learning Target:
Understand numbers.

Chapter Success Criteria:
- ■ I can identify a group of objects.
- ■ I can describe numbers as a group.
- ■ I can write numbers.
- ■ I can count objects.

8 Vocabulary

© Big Ideas Learning, LLC

Review Words
five
addition sentence

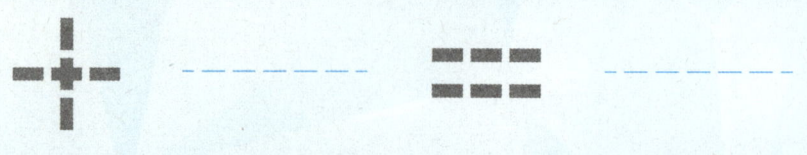

Directions: Count the shooting stars in the sky. Write the number. You see 2 more shooting stars in the sky. Draw the shooting stars. Then write an addition sentence to tell how many shooting stars there are in all.

Chapter 8 Vocabulary Cards

eighteen

eleven

fifteen

fourteen

nineteen

seventeen

sixteen

thirteen

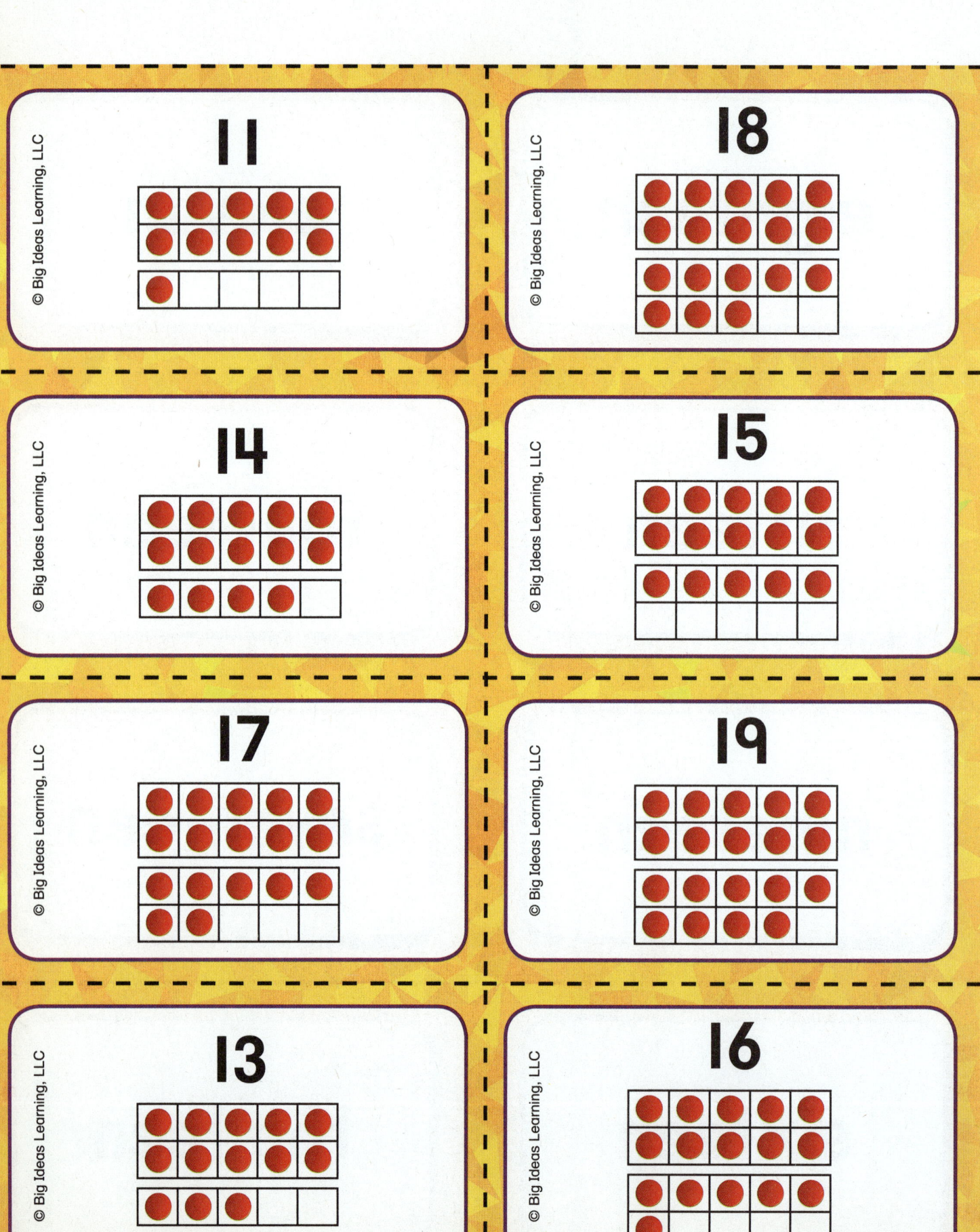

Chapter 8 Vocabulary Cards

twelve

12

Name _____

Learning Target: Find a group of 10 objects and tell how many more objects there are.

 **Explore and Grow**

Directions: Count and circle 10 linking cubes. Color the extra linking cubes.

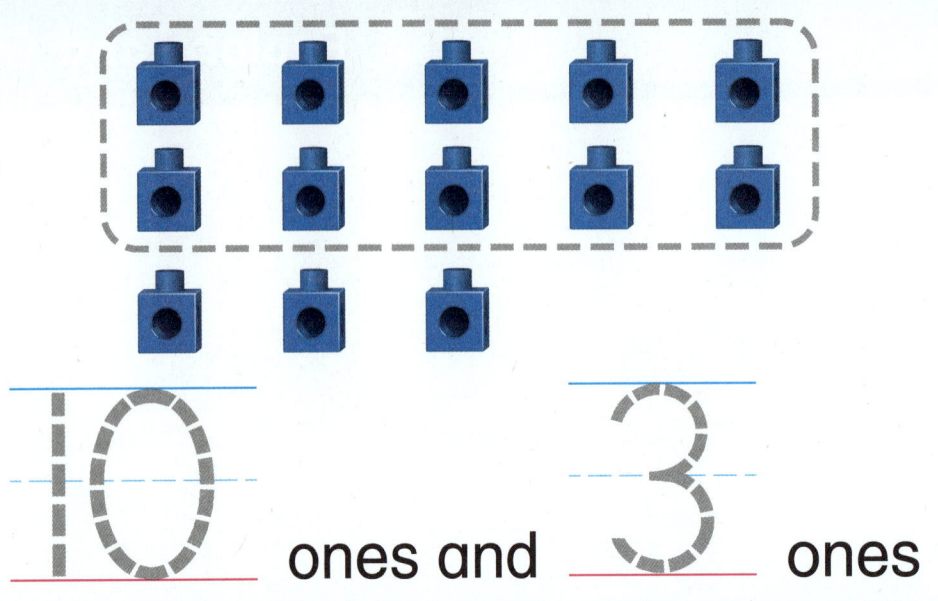

10 ones and 3 ones

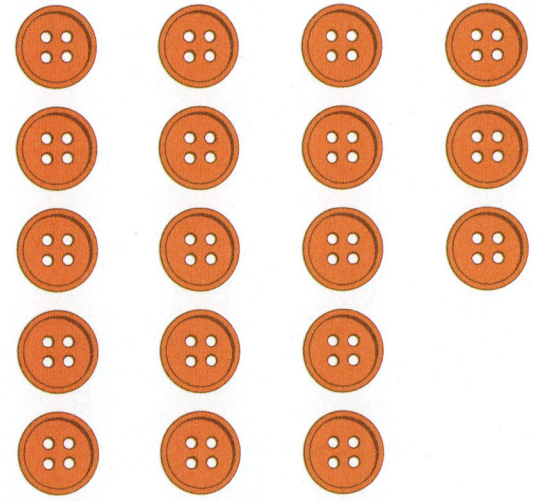

10 ones and _____ ones

Directions: Circle 10 objects. Tell how many more objects there are. Then write the numbers.

✔ Apply and Grow: Practice

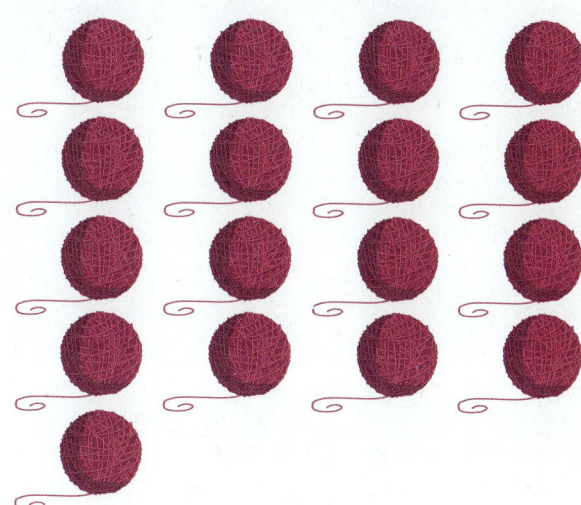

10 _____ ones and _____ ones

_____ ones and _____ ones

Directions: 1 and 2 Circle 10 objects. Tell how many more objects there are. Then write the numbers.

10 ones and **1** one

10 ones and **6** ones

Directions: Draw beads on the string to show how many beads there are in all.
Circle 10 beads.

Name _____

Learning Target: Find a group of 10 objects and tell how many more objects there are.

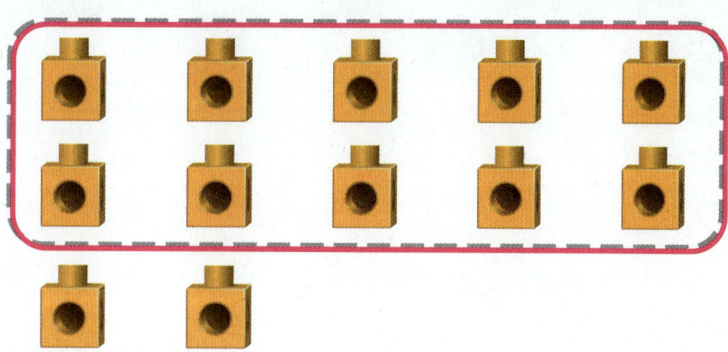

 ones and 2 ones

Directions: Circle 10 linking cubes. Tell how many more linking cubes there are. Then write the numbers.

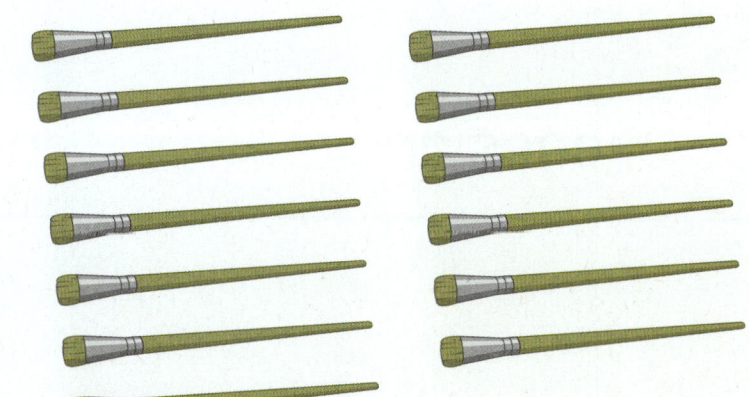

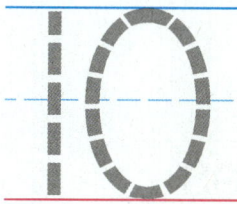 ones and _____ ones

Directions: Circle 10 paintbrushes. Tell how many more paintbrushes there are. Then write the numbers.

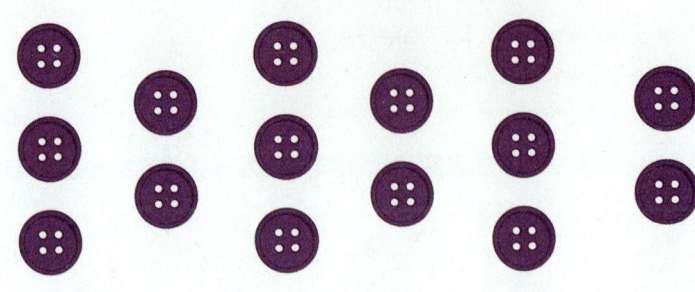

_____

- - - - - - - - -

_____ ones and _____ ones

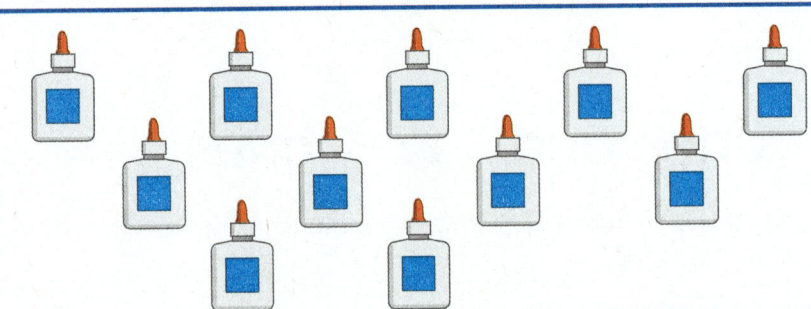

_____

- - - - - - - - -

_____ ones and _____ ones

10 ones and 7 ones

Directions: and Circle 10 objects. Tell how many more objects there are. Then write the numbers. Draw beads on the string to show how many beads there are in all. Circle 10 beads.

384   three hundred eighty-four

Name _____

Learning Target: Count and
write the numbers 11 and 12.

 Explore and Grow

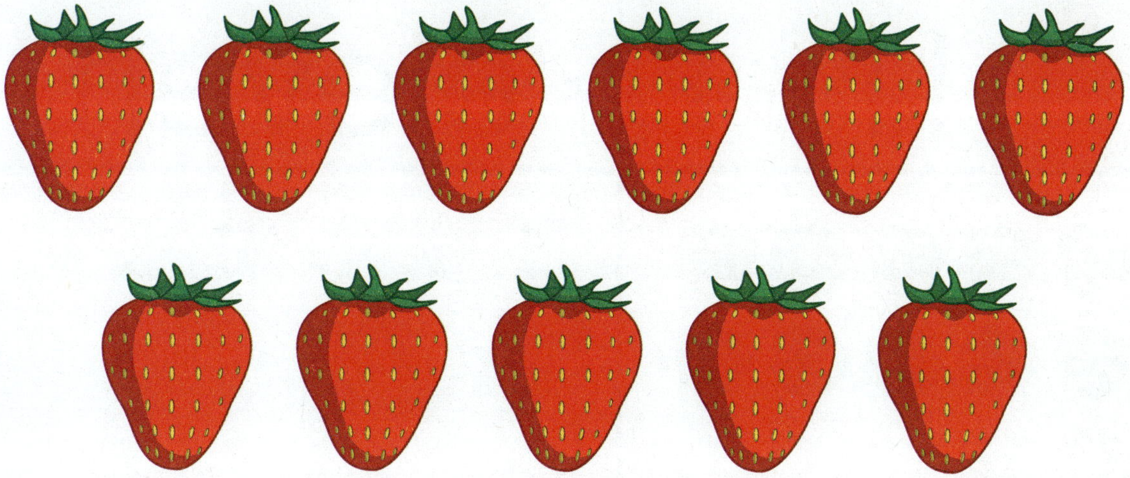

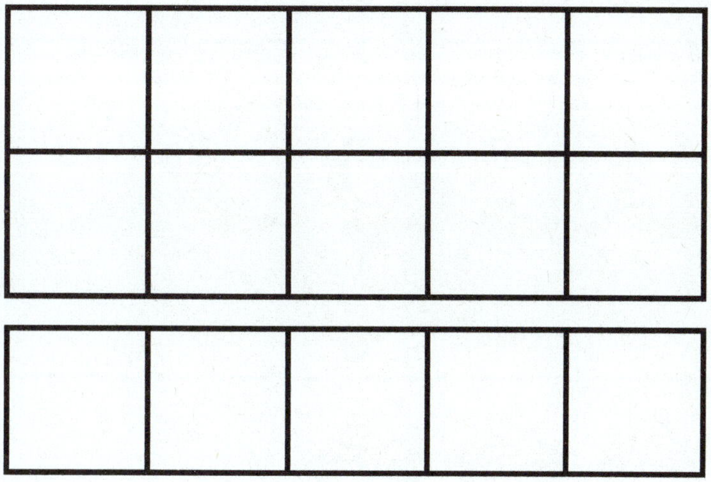

Directions: Place a linking cube on each strawberry. Slide cubes to fill the ten frame. Slide the extra cubes to the five frame.

© Big Ideas Learning, LLC

11

eleven

12

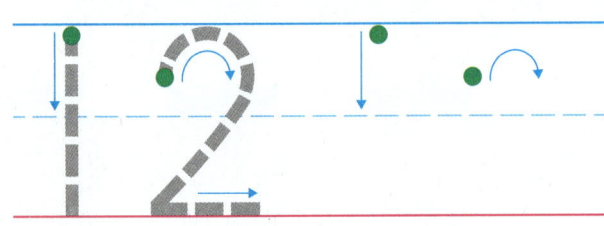

twelve

Directions:
- Count the fruit. Say the number. Trace and write the number.
- Count the fruit. Say the number. Write the number.

Name _____

 1

_____
- - - - - - - - -
_____

2

_____
- - - - - - - - -
_____

3

_____
- - - - - - - - -
_____

4

_____
- - - - - - - - -
_____

Directions: ● – ● Count the objects. Say the number. Write the number.

Chapter 8 | Lesson 2

Think and Grow: Modeling Real Life

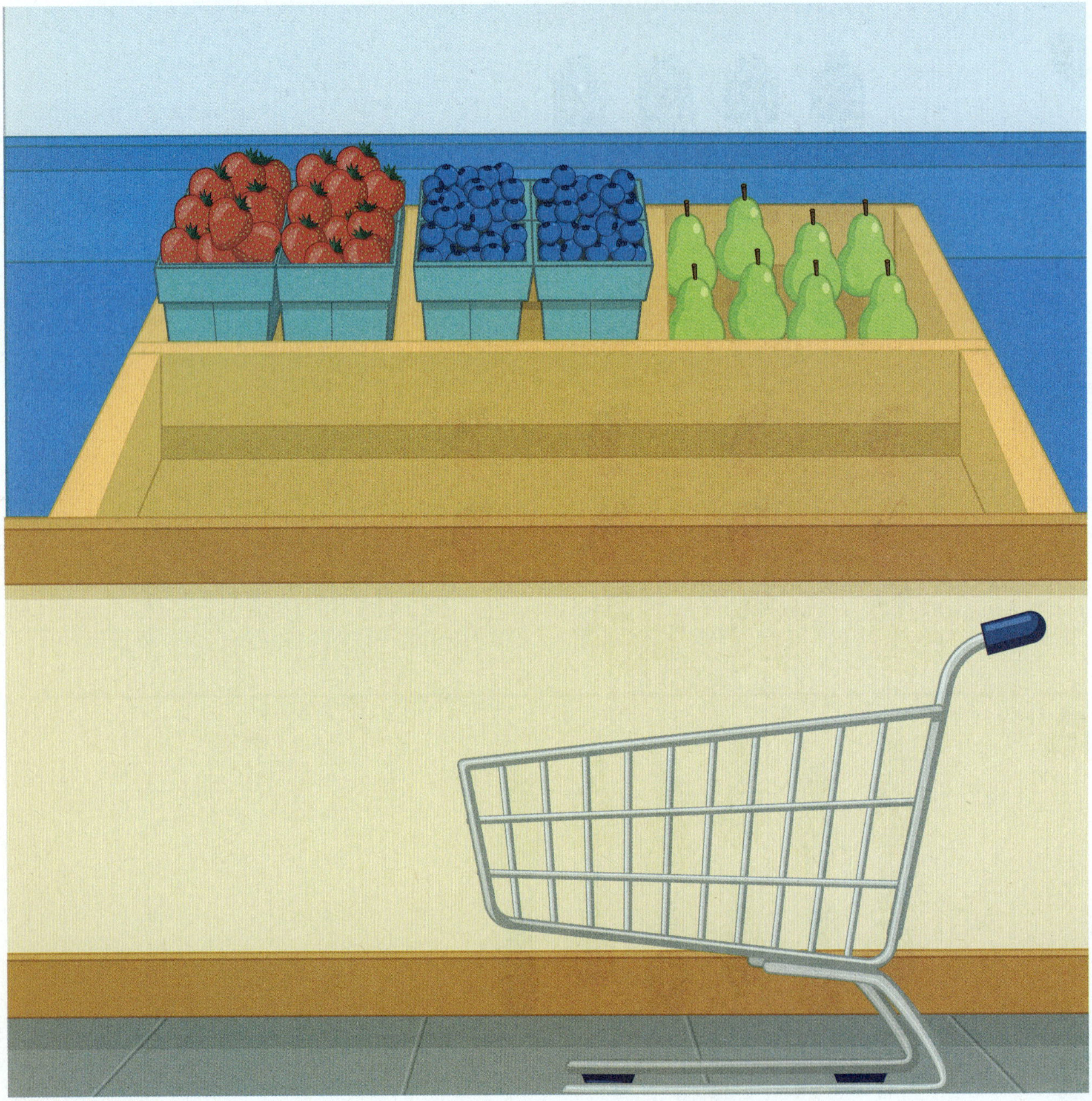

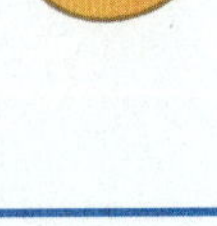

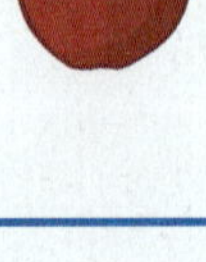

Directions: You have 12 oranges in your cart. There are 11 apples in the bin. Draw the oranges in the cart and the apples in the bin. Then write the numbers.

Learning Target: Count and write the numbers 11 and 12.

eleven

twelve

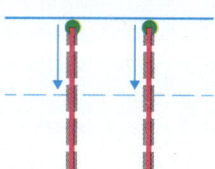 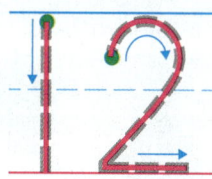

Directions: Count the linking cubes. Say the number. Write the number.

 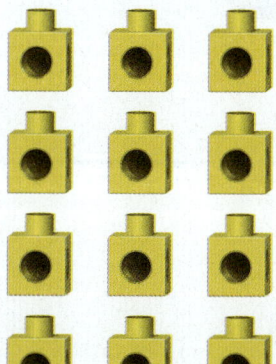

Directions: 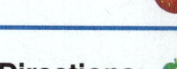 and  Count the objects. Say the number. Write the number.

3

_____

- - - - - - - - - - -

_____

4

_____

- - - - - - - - - - -

_____

5

_____

- - - - - - - - - - -

_____

Directions: **3** and **4** Count the fruit. Say the number. Write the number.
5 Draw 12 cherries on the tree. Write the number.

Learning Target: Understand the numbers 11 and 12.

 **Explore and Grow**

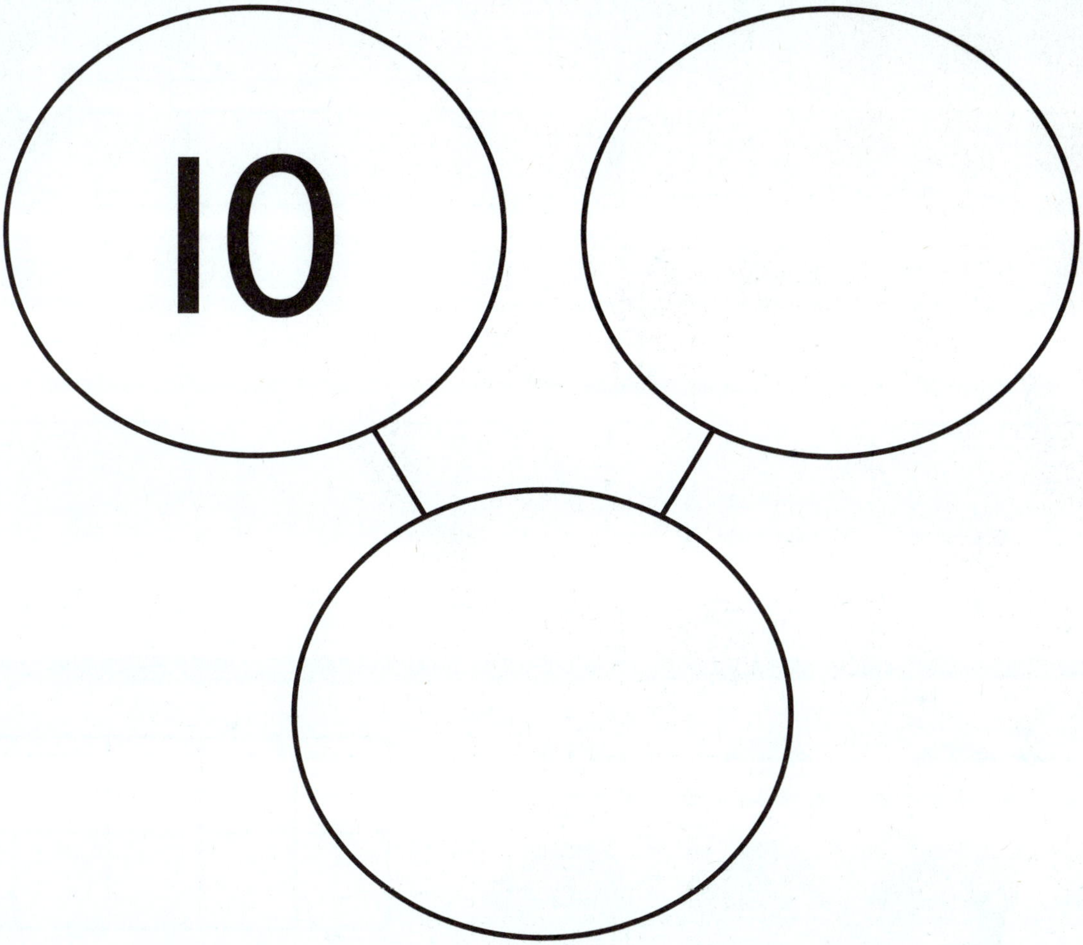

10

10 ___ and ___ is ___.

Directions: Place the 10 and 1 cards as the parts on the number bond. Slide the cards and hide the zero with the 1 card to make the whole. Write the parts and the whole.

💭 10 ones

💭 I one

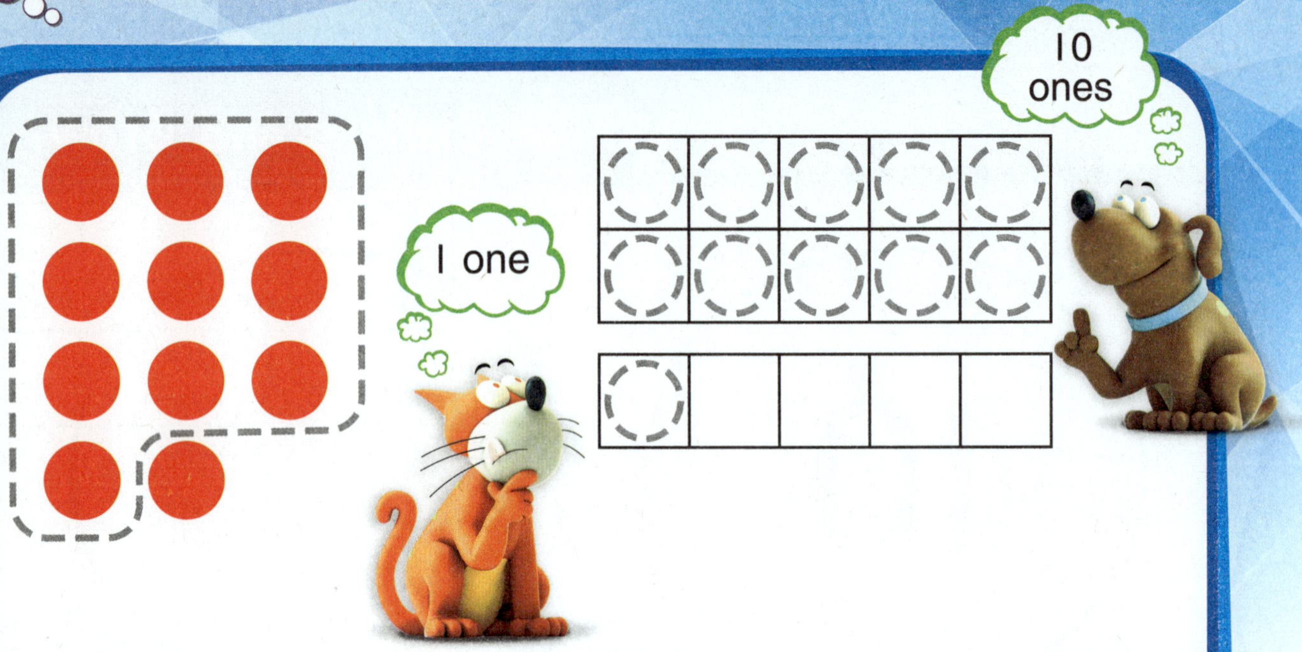

11 = 10 + 1

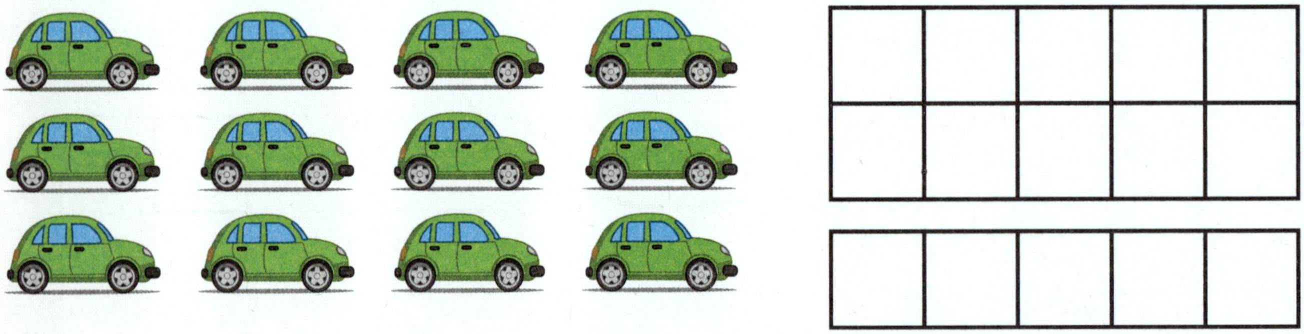

___ = 10 + ___

Directions: Circle 10 objects. Draw dots in the ten frame to show how many objects are circled. Draw dots in the five frame to show how many more objects there are. Use the frames to write an addition sentence.

✔ Apply and Grow: Practice

 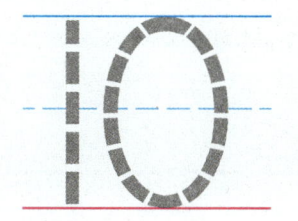 

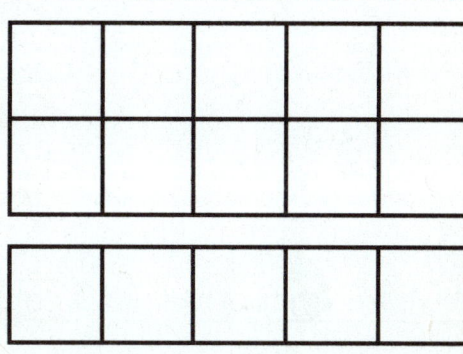

Directions: 1 and 2 Circle 10 vehicles. Draw dots in the ten frame to show how many vehicles are circled. Draw dots in the five frame to show how many more vehicles there are. Use the frames to write an addition sentence.

 # Think and Grow: Modeling Real Life

_____ ✛ _ _ _ _ _ _ _ _ _ ═ _____

_____ ✛ _ _ _ _ _ _ _ _ _ ═ _____

Directions:

- You have blue trucks. Your friend has red trucks. Circle your trucks. Write an addition sentence to match the picture. How many trucks do you have? Circle the number.

- You have yellow trains. Your friend has green trains. Circle your trains. Write an addition sentence to match the picture. How many trains does your friend have? Circle the number.

Learning Target: Understand the numbers 11 and 12.

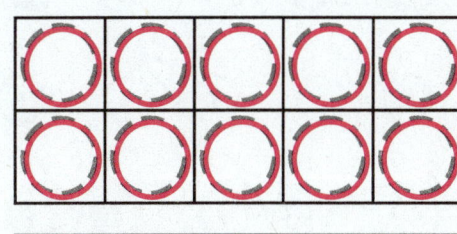
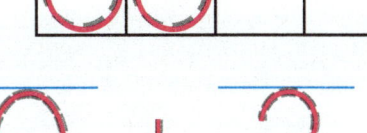

12 = 10 + 2

Directions: Circle 10 objects. Draw dots in the ten frame to show how many objects are circled. Draw dots in the five frame to show how many more objects there are. Use the frames to complete the addition sentence.

 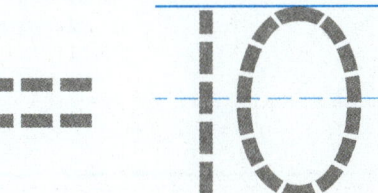 

== 10 +

Directions: ① Circle 10 trucks. Draw dots in the ten frame to show how many trucks are circled. Draw dots in the five frame to show how many more trucks there are. Use the frames to write an addition sentence.

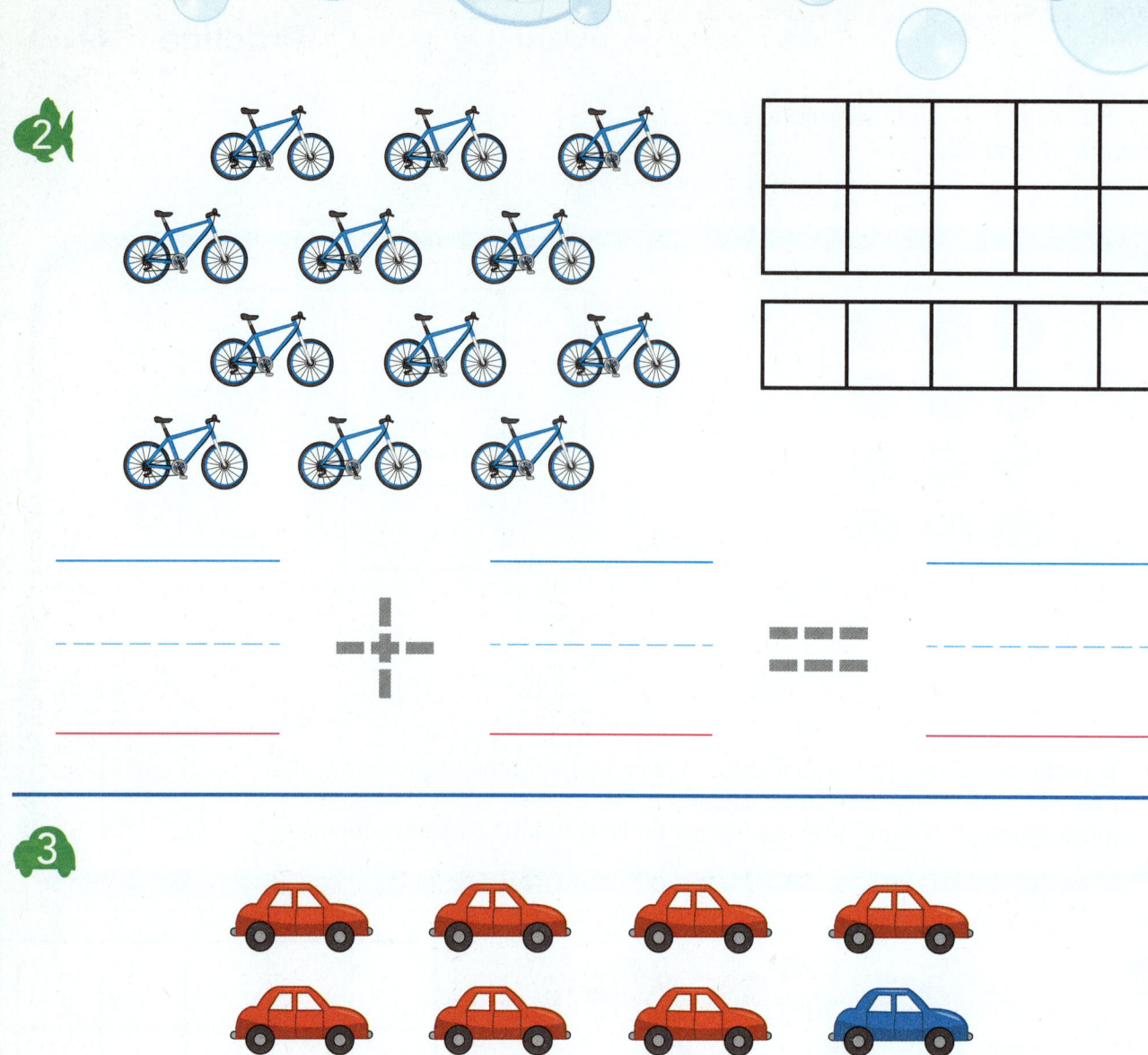

2

_____ + _____ = _____

3

_____ + _____ = _____

Directions: **2** Circle 10 bicycles. Draw dots in the ten frame to show how many bicycles are circled. Draw dots in the five frame to show how many more bicycles there are. Use the frames to write an addition sentence. **3** You have red cars. Your friend has blue cars. Circle your cars. Write an addition sentence to match your picture. How many cars does your friend have? Circle the number.

Learning Target: Count and
write the numbers 13 and 14.

 **Explore and Grow**

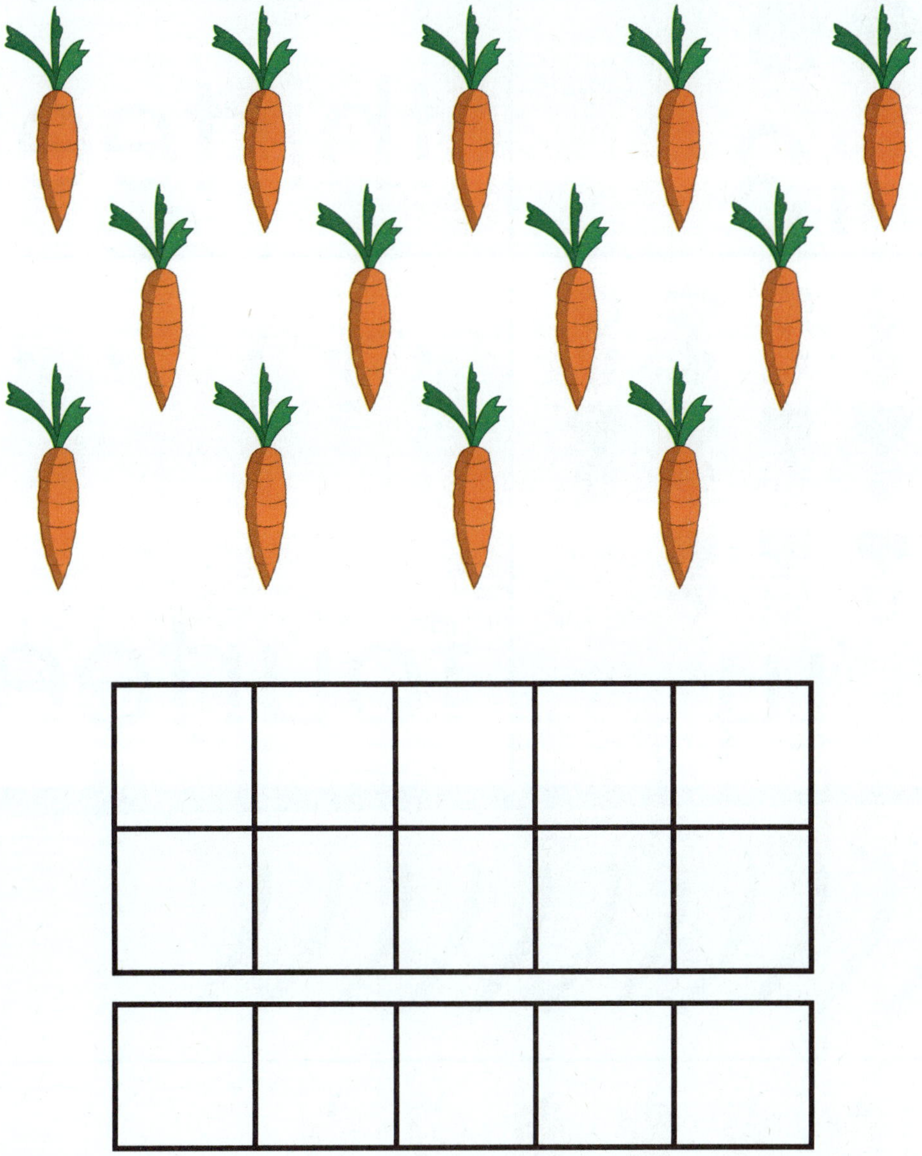

Directions: Place a linking cube on each carrot. Slide cubes to fill the ten frame.
Slide the extra cubes to the five frame.

13

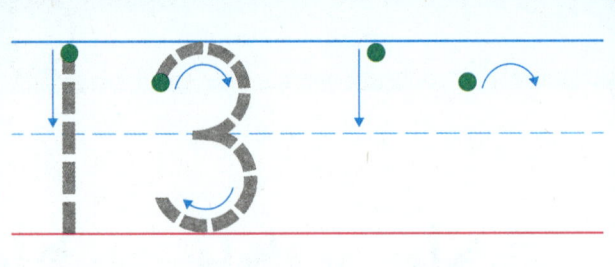

thirteen

14

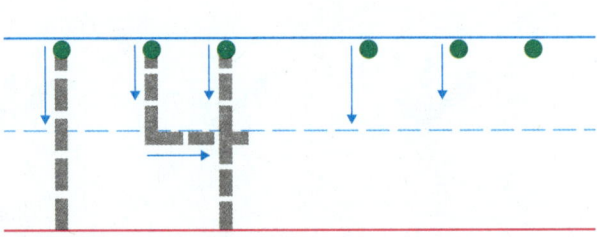

fourteen

Directions:
- Count the vegetables. Say the number. Trace and write the number.
- Count the vegetables. Say the number. Write the number.

Name _____

Directions: 🍎–🐸 Count the objects. Say the number. Write the number.

Think and Grow: Modeling Real Life

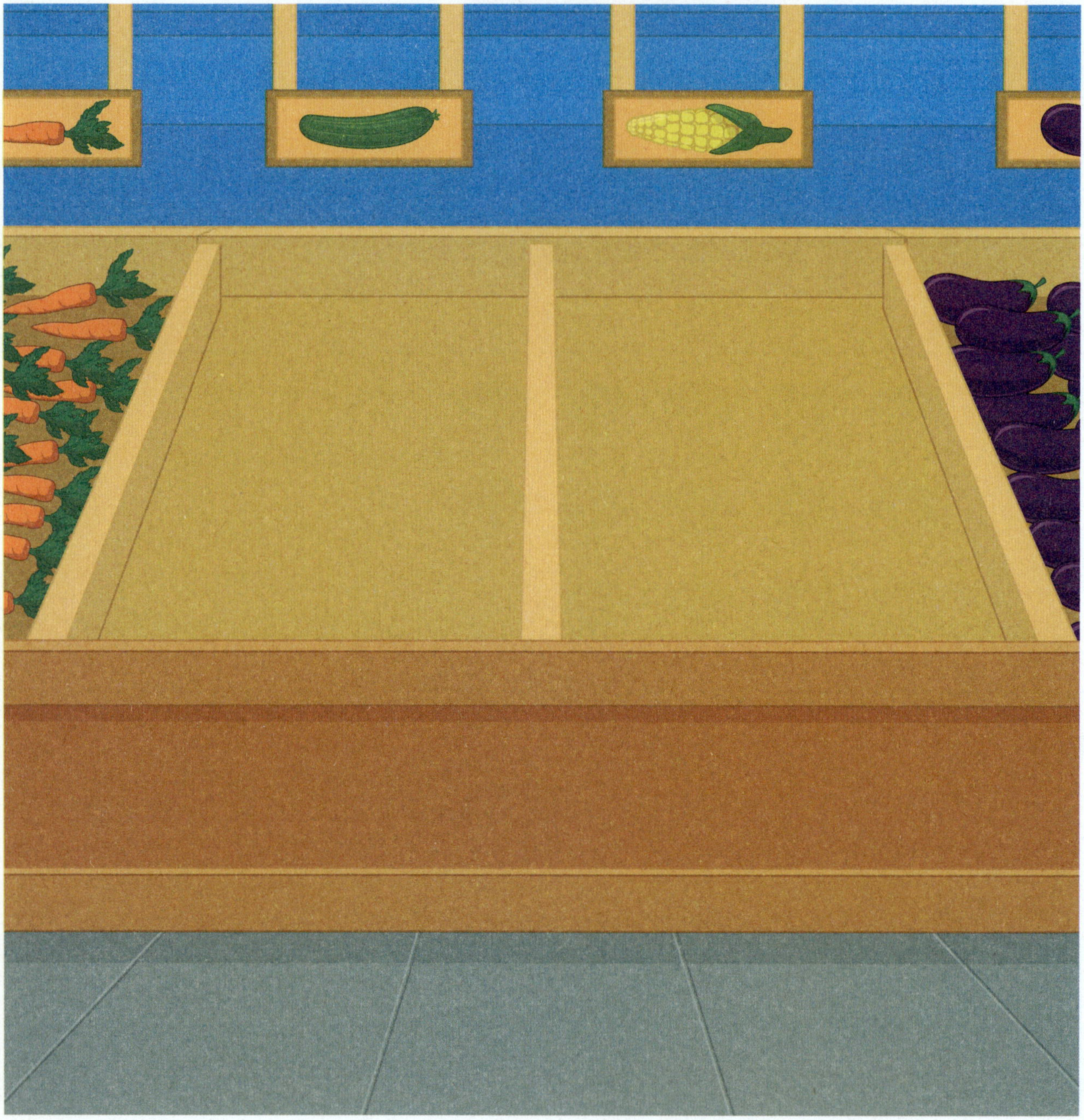

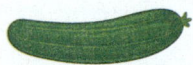

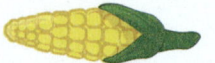

Directions: A store has 14 cucumbers and 13 ears of corn in the bins. Draw the cucumbers and the ears of corn. Then write the numbers.

Learning Target: Count and write the numbers 13 and 14.

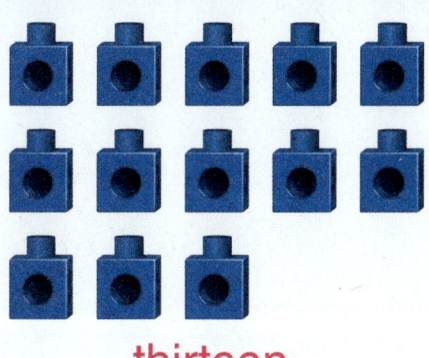

thirteen

fourteen

Directions: Count the linking cubes. Say the number. Write the number.

_____

_____

_____

_____

_____

_____

Directions: 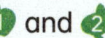 and ❷ Count the objects. Say the number. Write the number.

3

_____
- - - - - - - - -
_____

 4

_____
- - - - - - - - -
_____

 5

_____
- - - - - - - - -
_____

Directions: and Count the vegetables. Say the number. Write the number.
 Draw 14 heads of lettuce in the dirt. Write the number.

© Big Ideas Learning, LLC

402   four hundred two

Learning Target: Understand the numbers 13 and 14.

 **Explore and Grow**

10

___ and ___ is ___ .

Directions: Place the 10 and 3 cards as the parts on the number bond. Slide the cards and hide the zero with the 3 card to make the whole. Write the parts and the whole.

© Big Ideas Learning, LLC

10 ones

3 ones

13 = 10 + 3

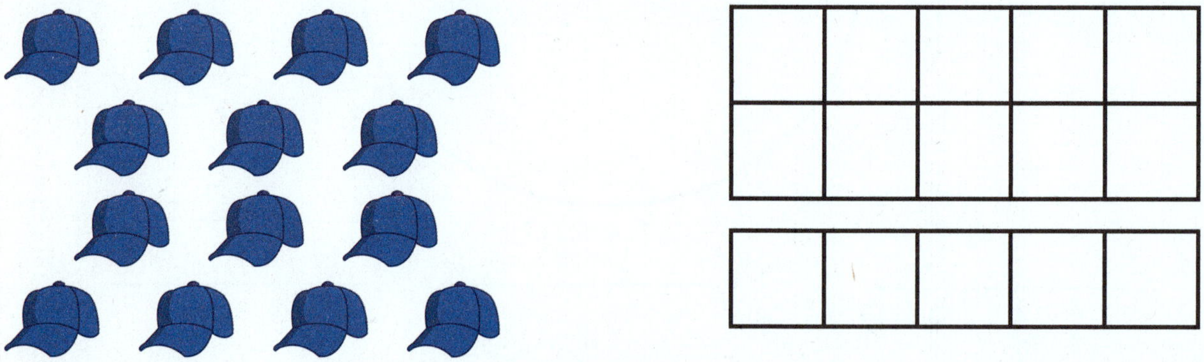

____ = 10 + ____

© Big Ideas Learning, LLC

Directions: Circle 10 objects. Draw dots in the ten frame to show how many objects are circled. Draw dots in the five frame to show how many more objects there are. Use the frames to write an addition sentence.

✔ Apply and Grow: Practice

1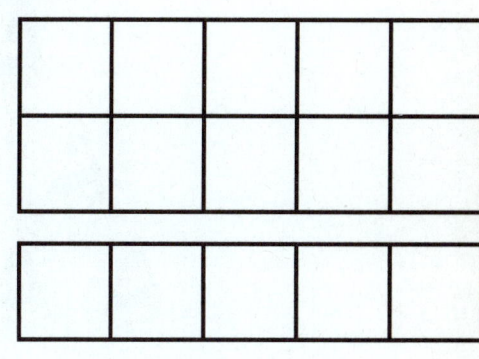

2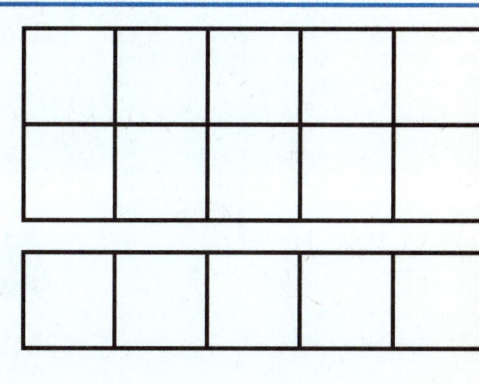

Directions: **1** and **2** Circle 10 hats. Draw dots in the ten frame to show how many hats are circled. Draw dots in the five frame to show how many more hats there are. Use the frames to write an addition sentence.

 # Think and Grow: Modeling Real Life

_____   **+**   _____   **=**   _____

_____   **+**   _____   **=**   _____

Directions:
- You have striped party hats. Your friend has polka-dot party hats. Circle your hats. Write an addition sentence to match the picture. How many hats do you and your friend have in all? Circle the number.
- You have red party hats. Your friend has blue party hats. Circle your hats. Write an addition sentence to match the picture. How many hats do you have? Circle the number.

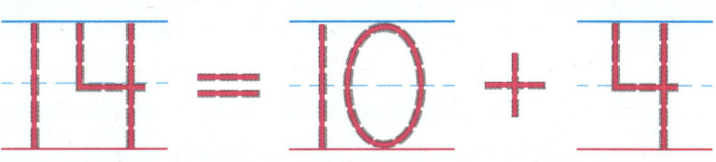

Directions: Circle 10 objects. Draw dots in the ten frame to show how many objects are circled. Draw dots in the five frame to show how many more objects there are. Use the frames to complete the addition sentence.

1

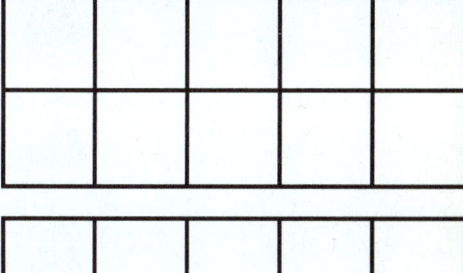

Directions: **1** Circle 10 hats. Draw dots in the ten frame to show how many hats are circled. Draw dots in the five frame to show how many more hats there are. Use the frames to write an addition sentence.

2

3

Directions: **2** Circle 10 hats. Draw dots in the ten frame to show how many hats are circled. Draw dots in the five frame to show how many more hats there are. Use the frames to write an addition sentence. **3** You have green party hats. Your friend has purple party hats. Circle your hats. Write an addition sentence to match the picture. How many hats do you have? Circle the number.

Name ______________________

Count and Write 15 — 8.6

Learning Target: Count and write the number 15.

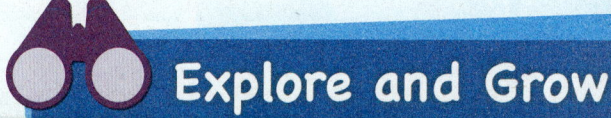

Explore and Grow

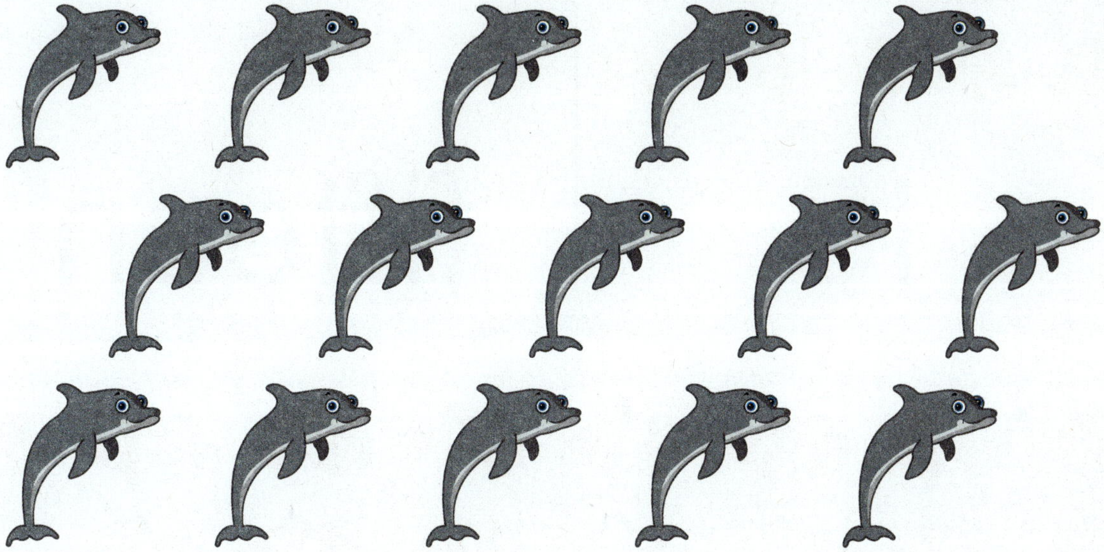

Directions: Place a linking cube on each dolphin. Slide cubes to fill the top ten frame. Slide the extra cubes to the bottom ten frame.

Think and Grow

15

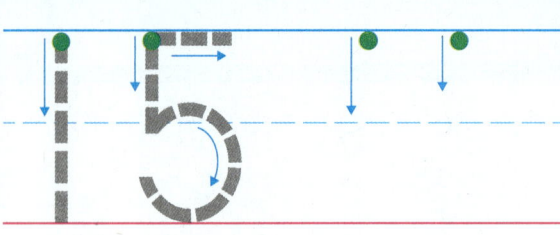

fifteen

Directions:
- Count the fish. Say the number. Trace and write the number.
- Count the fish. Say the number. Write the number.

Name _____

_____
- - - - - - - - -
_____

_____
- - - - - - - - -
_____

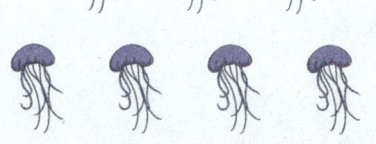

_____
- - - - - - - - -
_____

_____
- - - - - - - - -
_____

Directions: ❶–❹ Count the objects. Say the number. Write the number.

Directions: 11 blue fish swim near the top of the water. 15 red fish swim near the shipwreck. Draw the fish. Then write the numbers.

412   four hundred twelve

Name _____

Learning Target: Count and write the number 15.

fifteen

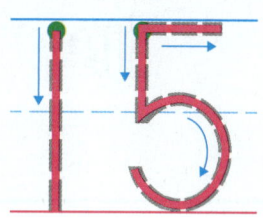

Directions: Count the linking cubes. Say the number. Write the number.

1

2

Directions: 1 and 2 Count the objects. Say the number. Write the number.

3

_____

- - - - - - - -

_____

4

_____

- - - - - - - -

_____

5

_____

- - - - - - - -

_____

Directions: **3** and **4** Count the sea creatures. Say the number. Write the number. **5** Draw 15 bubbles in the water. Write the number.

414   four hundred fourteen

Name _____

Learning Target: Understand the number 15.

 **Explore and Grow**

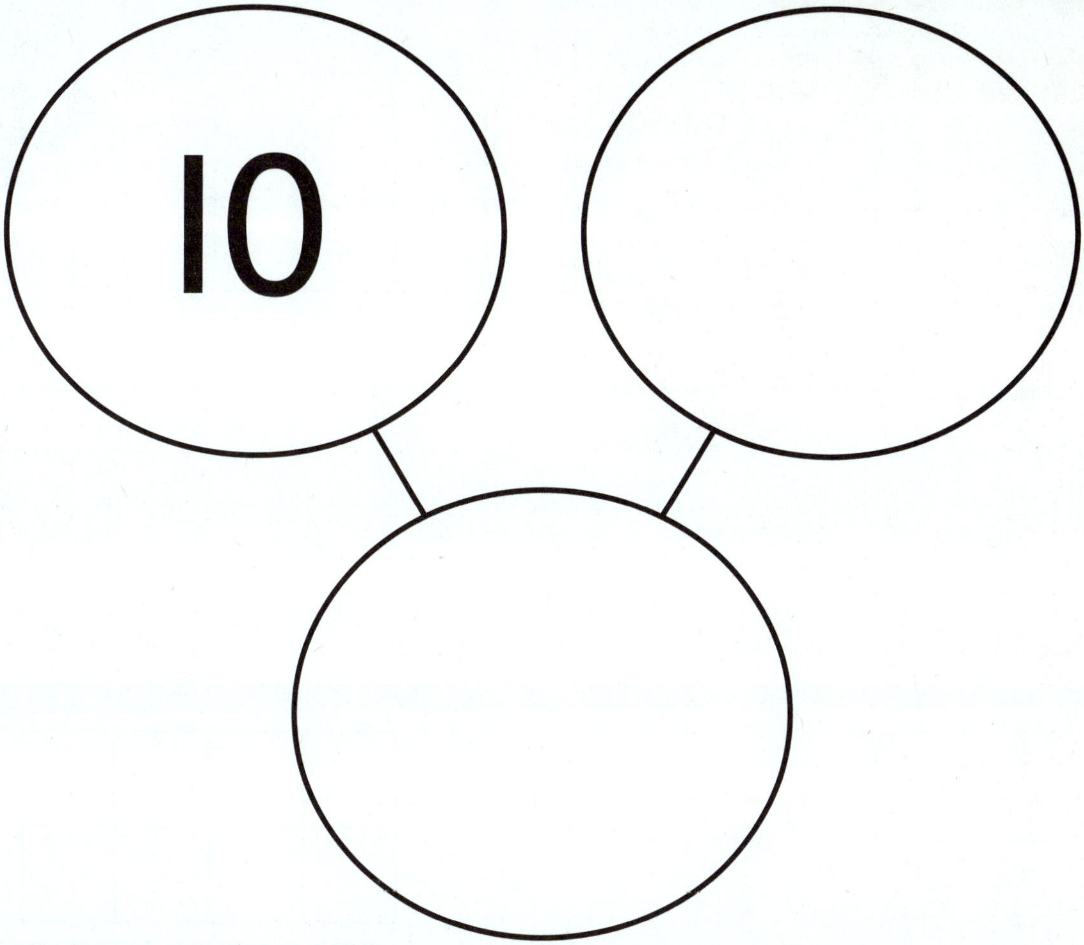

10

10 ___ and ___ is ___.

Directions: Place the 10 and 5 cards as the parts on the number bond. Slide the cards and hide the zero with the 5 card to make the whole. Write the parts and the whole.

Think and Grow

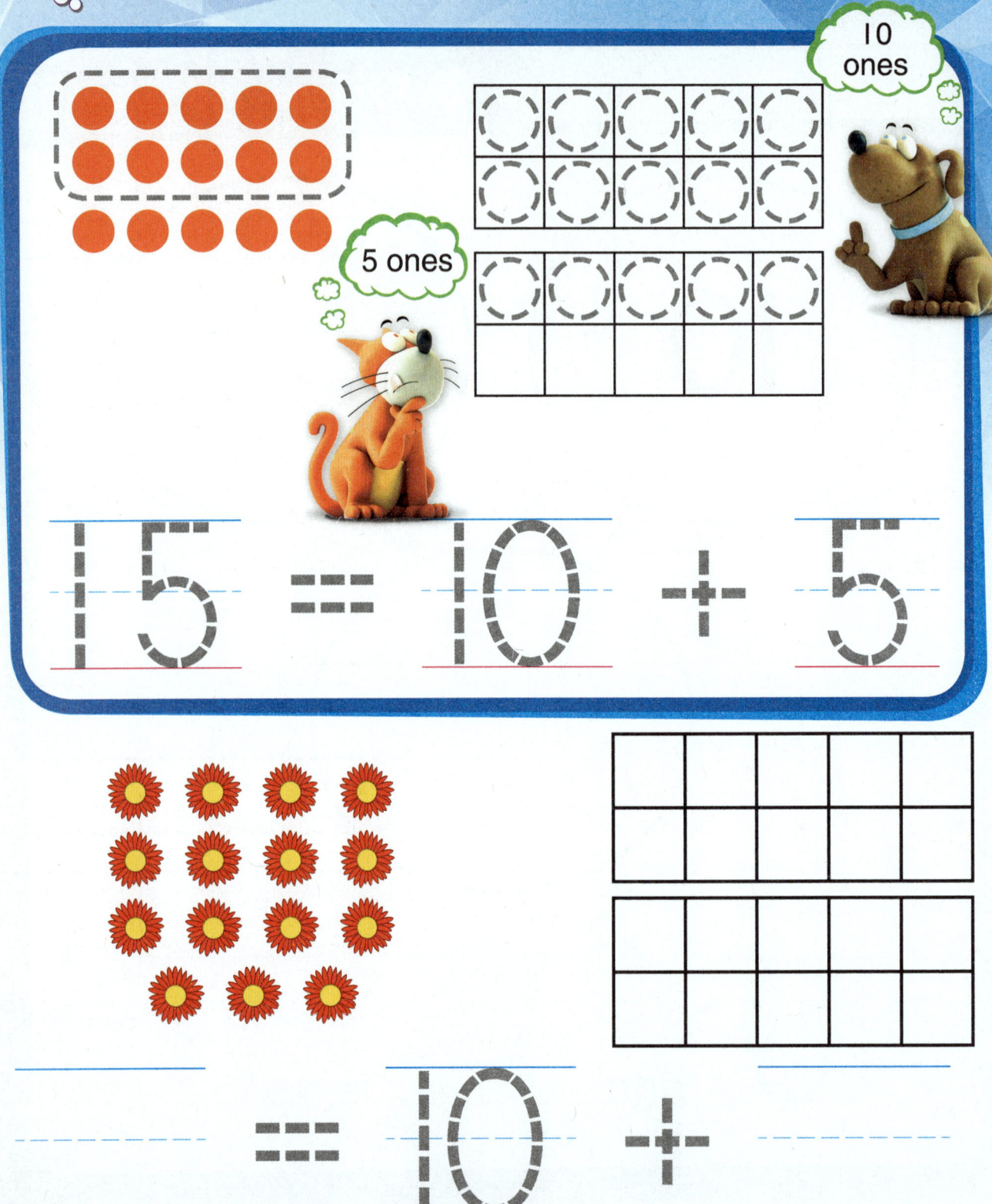

10 ones

5 ones

$$15 = 10 + 5$$

$$\underline{} = 10 + \underline{}$$

Directions: Circle 10 objects. Draw dots in the top ten frame to show how many objects are circled. Draw dots in the bottom ten frame to show how many more objects there are. Use the ten frames to write an addition sentence.

✔ Apply and Grow: Practice

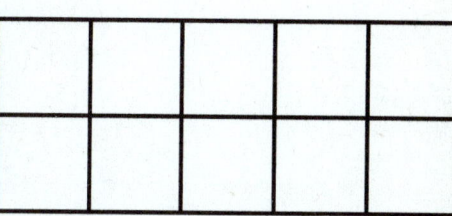

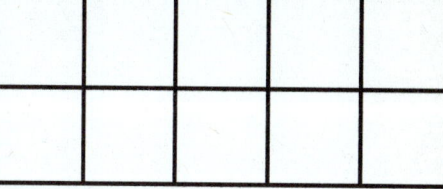

_____ = 10 + _____

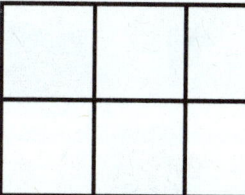

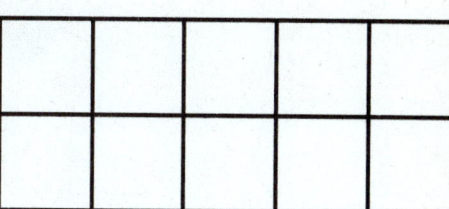

_____ = _____ + _____

Directions: 1️⃣ and 2️⃣ Circle 10 flowers. Draw dots in the top ten frame to show how many flowers are circled. Draw dots in the bottom ten frame to show how many more flowers there are. Use the ten frames to write an addition sentence.

_____ ╋ _ _ _ _ ═ _____

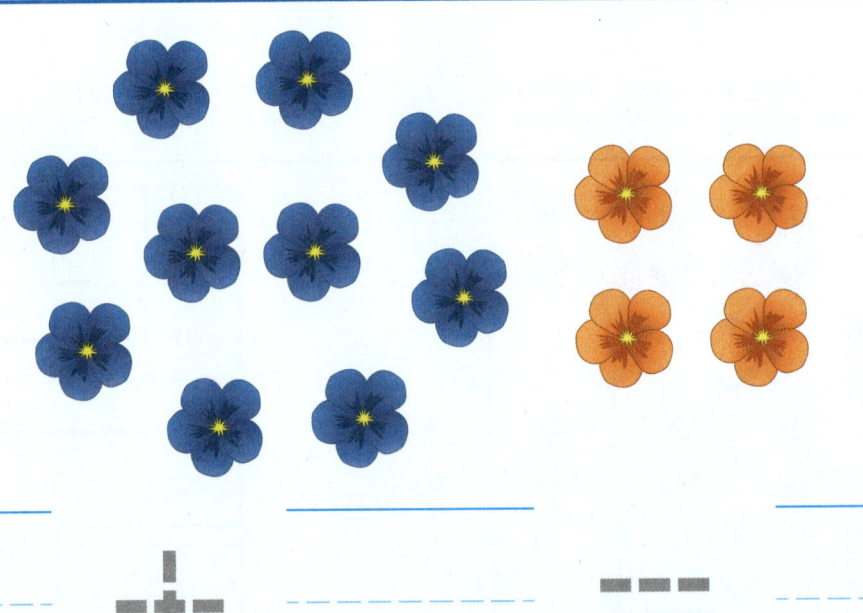

_____ ╋ _ _ _ _ ═ _____

Directions:

- You have pink flowers. Your friend has yellow flowers. Circle your flowers. Write an addition sentence to match the picture. How many flowers does your friend have? Circle the number.

- You have blue flowers. Your friend has orange flowers. Circle your flowers. Write an addition sentence to match the picture. How many flowers do you and your friend have in all? Circle the number.

Learning Target: Understand the number 15.

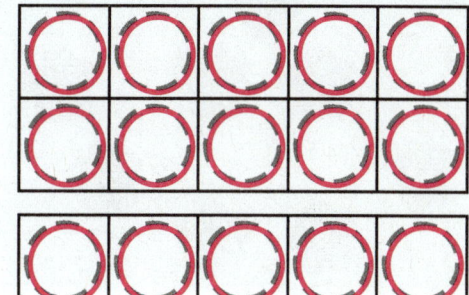

$$15 = 10 + 5$$

Directions: Circle 10 objects. Draw dots in the top ten frame to show how many objects are circled. Draw dots in the bottom ten frame to show how many objects are not circled. Use the ten frames to complete the addition sentence.

1

$$= 10 +$$

Directions: **1** Circle 10 flowers. Draw dots in the top ten frame to show how many flowers are circled. Draw dots in the bottom ten frame to show how many more flowers there are. Use the ten frames to write an addition sentence.

Directions: 2️⃣ Circle 10 flowers. Draw dots in the top ten frame to show how many flowers are circled. Draw dots in the bottom ten frame to show how many more flowers there are. Use the ten frames to write an addition sentence. 3️⃣ You have yellow flowers. Your friend has red flowers. Circle your flowers. Write an addition sentence to match the picture. How many flowers do you and your friend have in all? Circle the number.

Name _____

Learning Target: Count and
write the numbers 16 and 17.

 Explore and Grow

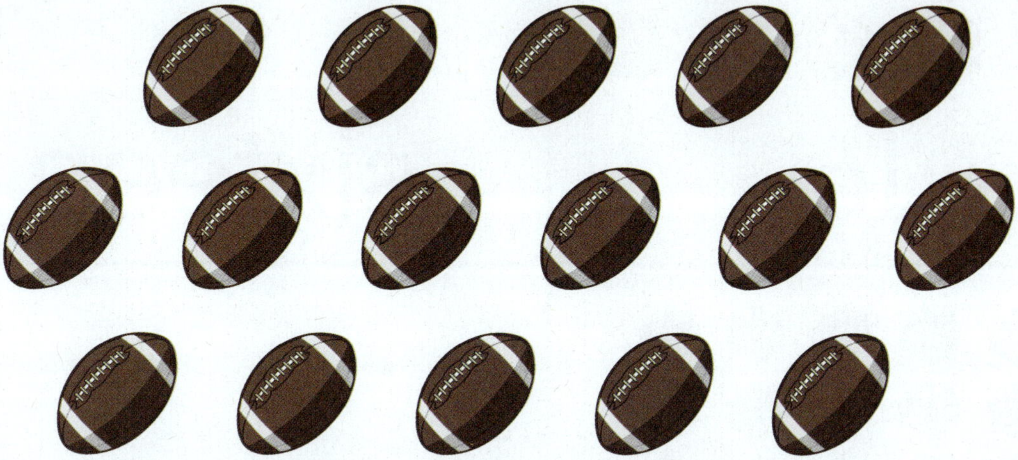

Directions: Place a linking cube on each football. Slide cubes to fill the top ten frame. Slide the extra cubes to the bottom ten frame.

© Big Ideas Learning, LLC

16

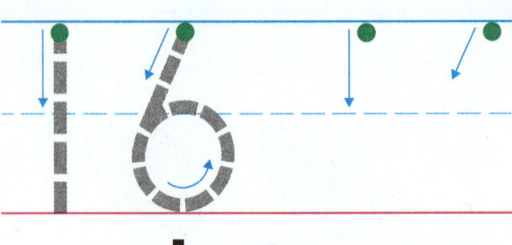

sixteen

17

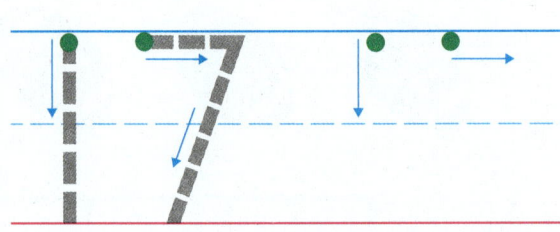

seventeen

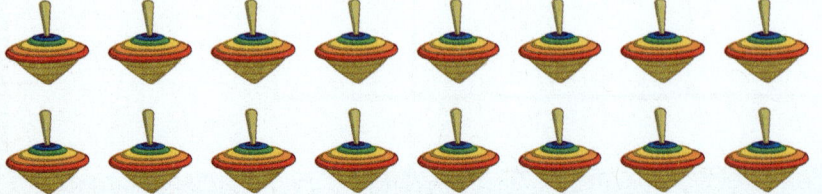

Directions:
- Count the objects. Say the number. Trace and write the number.
- Count the objects. Say the number. Write the number.

Name _____

 1

_____

- - - - - - - - - - - - -

_____

 **2**

_____

- - - - - - - - - - - - -

_____

 3

_____

- - - - - - - - - - - - -

_____

 **4**

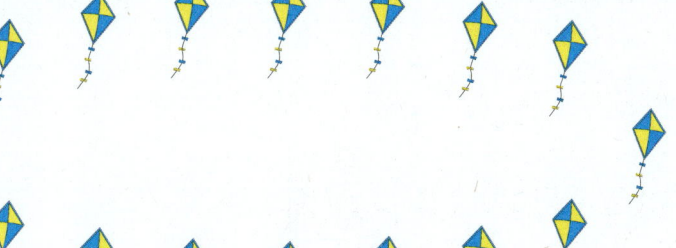

_____

- - - - - - - - - - - - -

_____

Directions:  – Count the objects. Say the number. Write the number.

Directions: There are 17 blocks on the floor and 16 books on the shelves. Draw the blocks and the books. Then write the numbers.

424   four hundred twenty-four

Learning Target: Count and write the numbers 16 and 17.

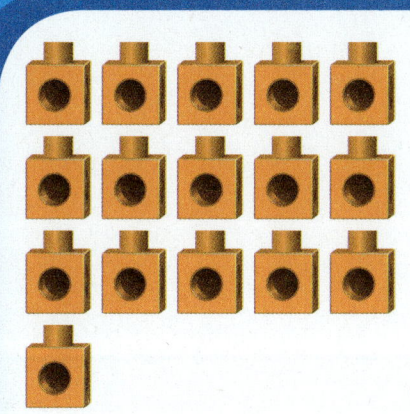 sixteen

seventeen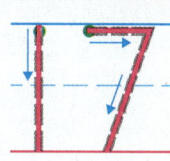

Directions: Count the linking cubes. Say the number. Write the number.

 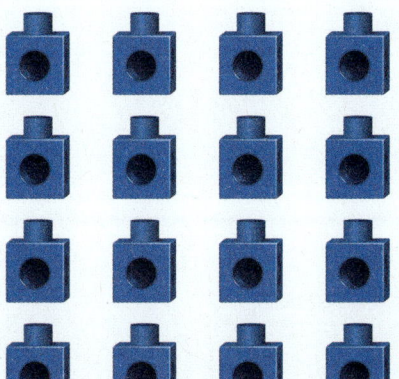

_____
- - - - - - - - - -
_____

  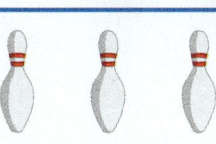  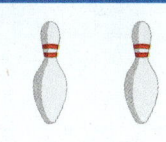 

_____
- - - - - - - - - -
_____

 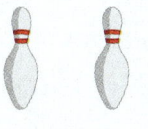 

  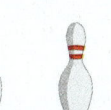 

Directions:  and ❷ Count the objects. Say the number. Write the number.

3

_____

- - - - - - - - - - - -

_____

4

_____

- - - - - - - - - - - -

_____

5

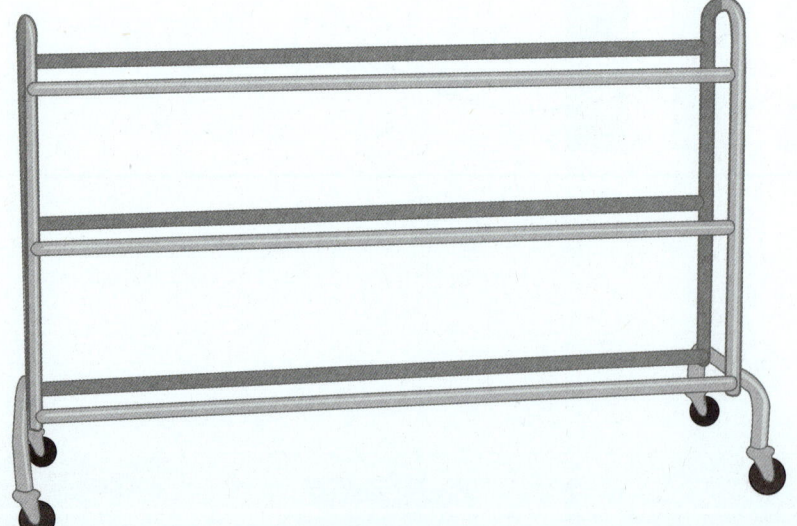

_____

- - - - - - - - - - - -

_____

Directions: **3** and **4** Count the objects. Say the number. Write the number.
5 Draw 17 balls on the ball rack. Write the number.

426   four hundred twenty-six

Name _____

Learning Target: Understand
the numbers 16 and 17.

Explore and Grow

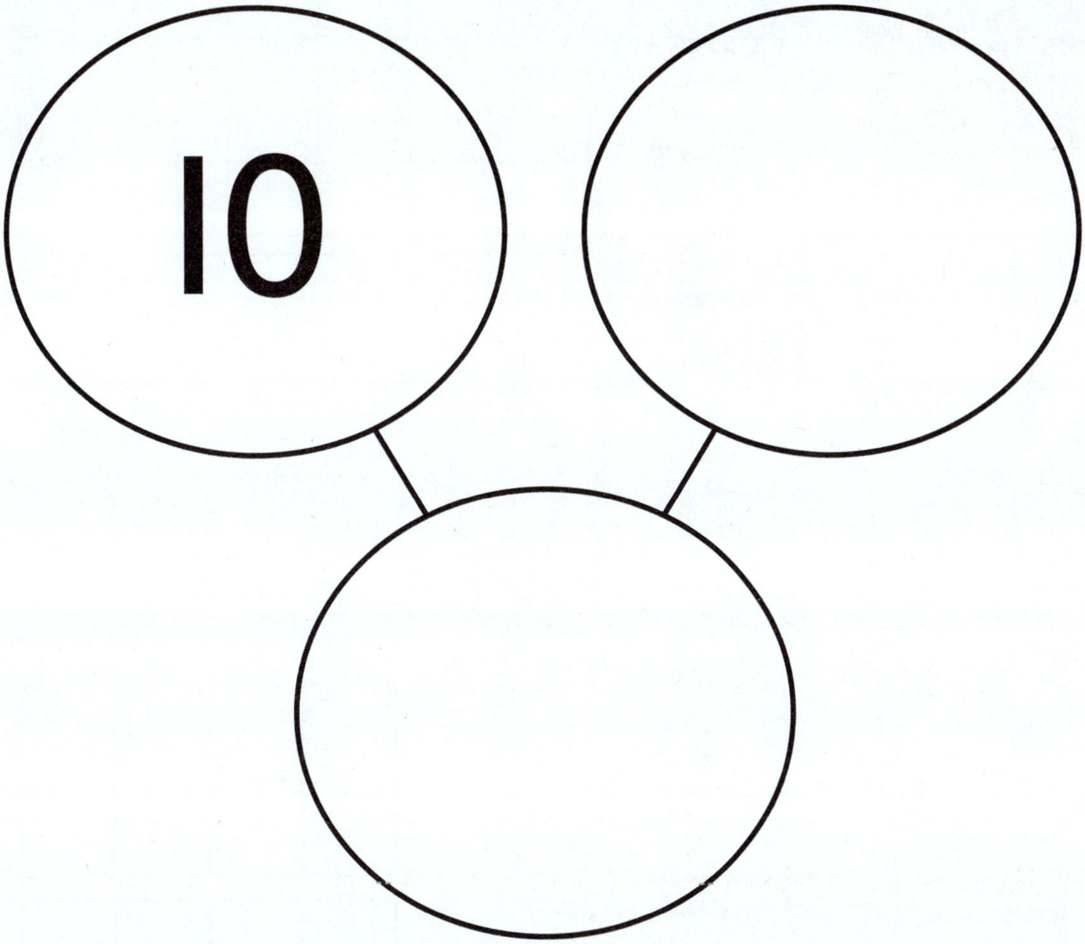

10

10 and _____ is _____ .

Directions: Place the 10 and 6 cards as the parts on the number bond. Slide the cards
and hide the zero with the 6 card to make the whole. Write the parts and the whole.

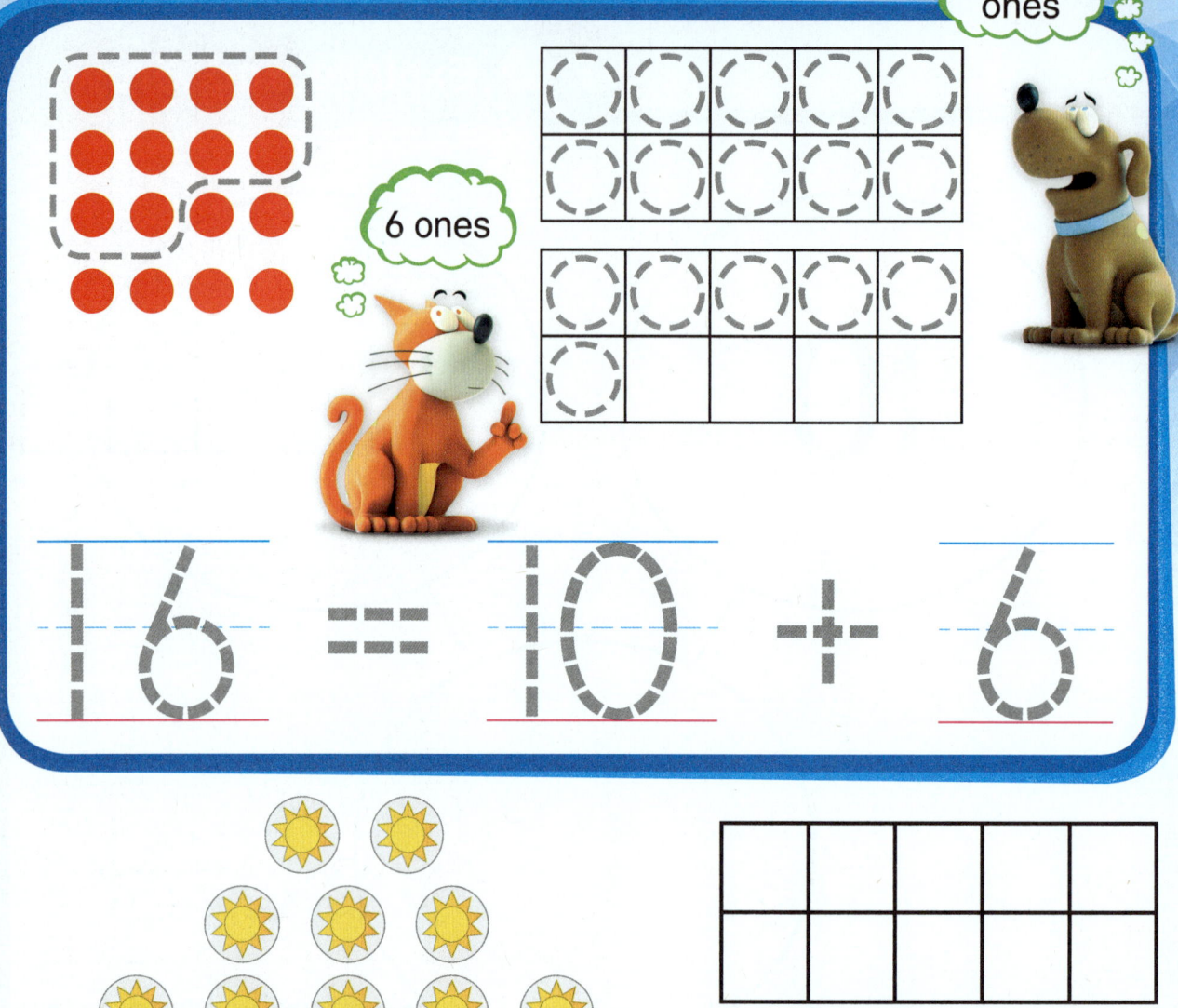

Directions: Circle 10 objects. Draw dots in the top ten frame to show how many objects are circled. Draw dots in the bottom ten frame to show how many more objects there are. Use the ten frames to write an addition sentence.

 **Apply and Grow: Practice**

1

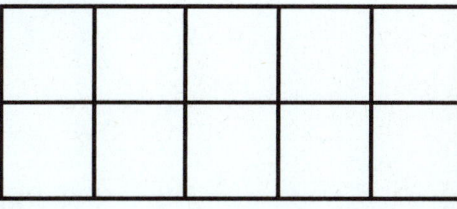

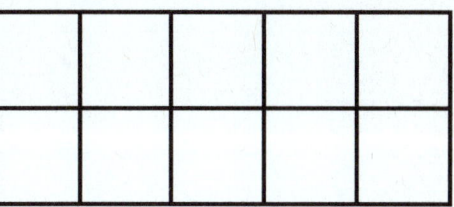

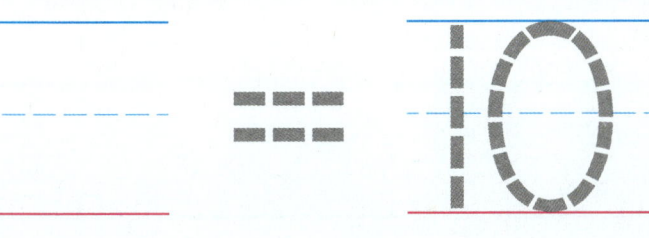

2

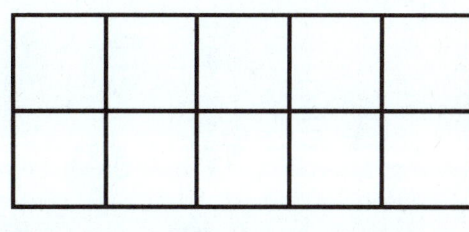

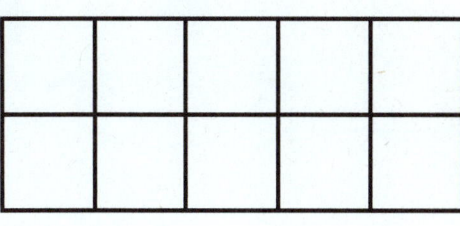

Directions: **1** and **2** Circle 10 objects. Draw dots in the top ten frame to show how many objects are circled. Draw dots in the bottom ten frame to show how many more objects there are. Use the ten frames to write an addition sentence.

Chapter 8 | Lesson 9

four hundred twenty-nine 429

Directions:

- You put yellow stickers on a page. Your friend puts red stickers on the same page. Circle your stickers. Write an addition sentence to match the picture. How many stickers does your friend put on the page? Circle the number.
- You put blue stickers on a page. Your friend puts green stickers on the same page. Circle your stickers. Write an addition sentence to match the picture. How many stickers does your friend put on the page? Circle the number.

Learning Target: Understand the numbers 16 and 17.

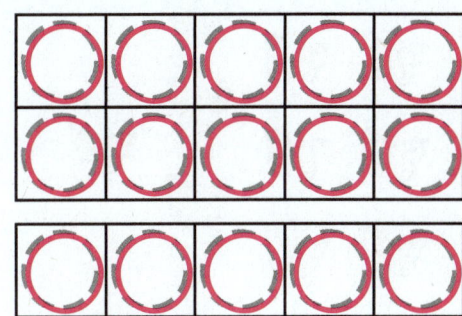

17 = 10 + 7

Directions: Circle 10 objects. Draw dots in the top ten frame to show how many objects are circled. Draw dots in the bottom ten frame to show how many objects are not circled. Use the ten frames to complete the addition sentence.

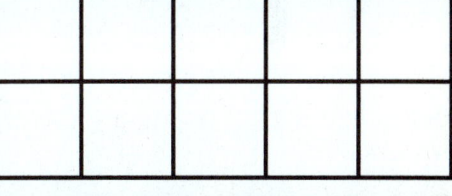
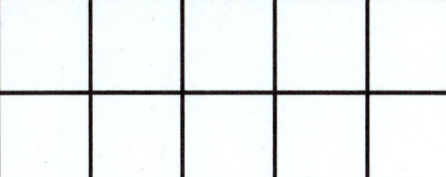

 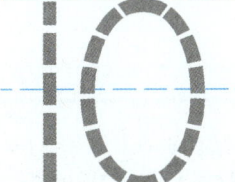 

_____ = 10 + _____

Directions: Circle 10 stars. Draw dots in the top ten frame to show how many stars are circled. Draw dots in the bottom ten frame to show how many more stars there are. Use the ten frames to write an addition sentence.

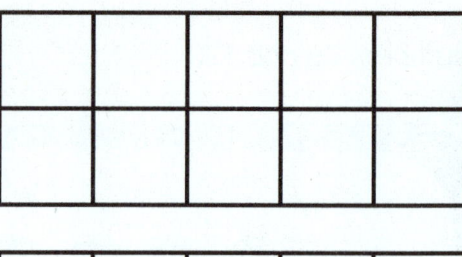

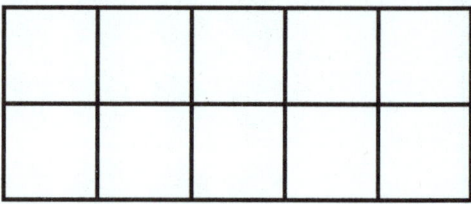

2

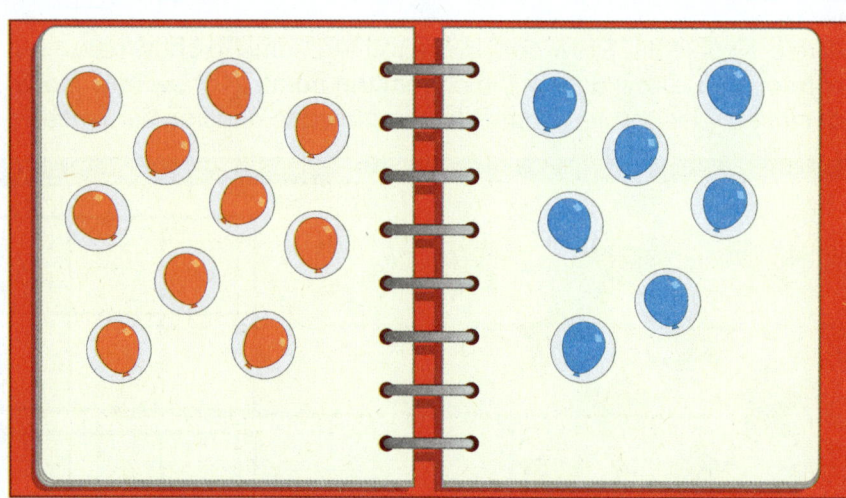

3

Directions: 2 Circle 10 butterflies. Draw dots in the top ten frame to show how many butterflies are circled. Draw dots in the bottom ten frame to show how many more butterflies there are. Use the ten frames to write an addition sentence. 3 You put orange stickers on a page. Your friend puts blue stickers on the next page. Circle your stickers. Write an addition sentence to match the picture. How many stickers does your friend put on the next page? Circle the number.

Learning Target: Count and
write the numbers 18 and 19.

Explore and Grow

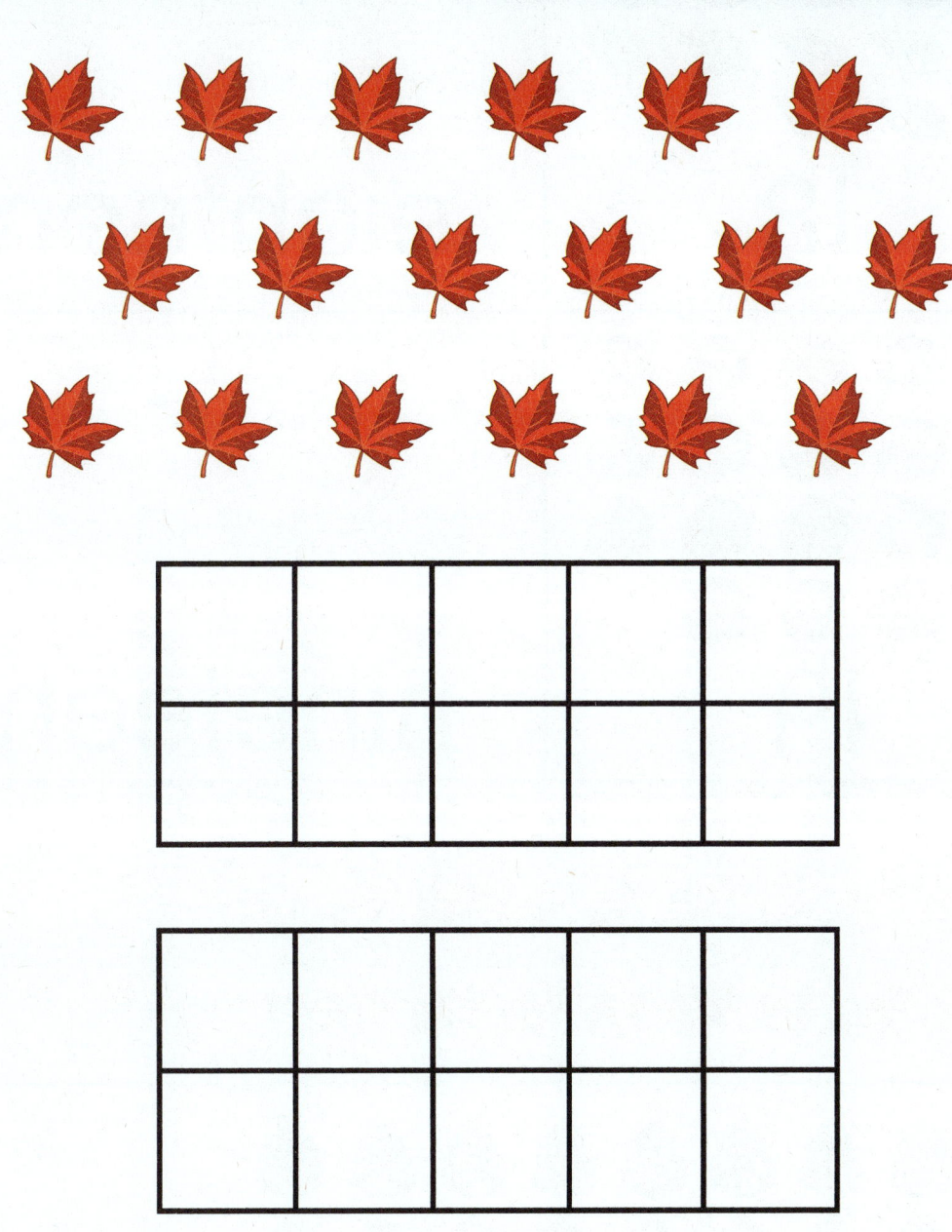

Directions: Place a linking cube on each leaf. Slide cubes to fill the top ten frame.
Slide the extra cubes to the bottom ten frame.

© Big Ideas Learning, LLC

18

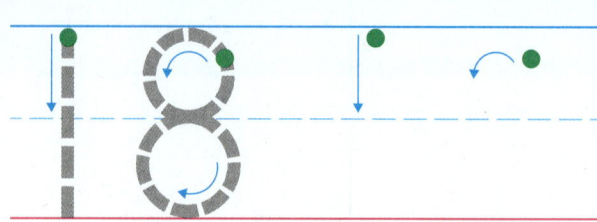

eighteen

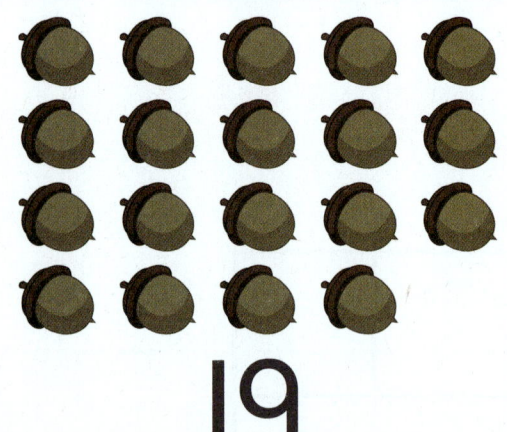

19

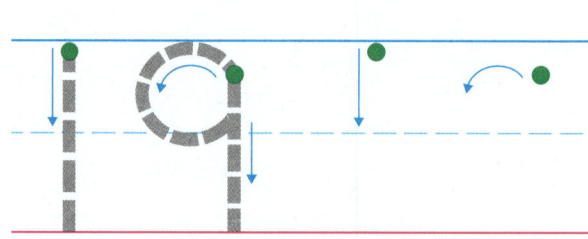

nineteen

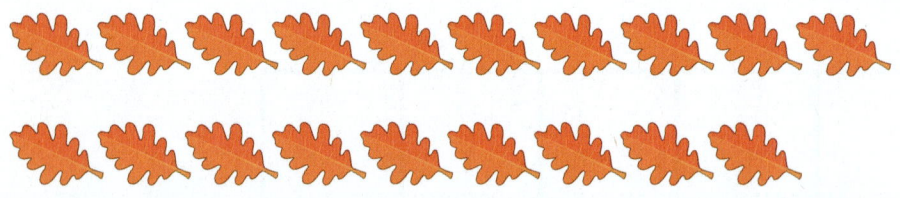

_____

- - - - - - - - - - -

_____

_____

- - - - - - - - - - -

_____

Directions:
- Count the objects. Say the number. Trace and write the number.
- Count the objects. Say the number. Write the number.

Name _____

 1

2

 3

 **4**

Directions: **1** – **4** Count the objects. Say the number. Write the number.

Think and Grow: Modeling Real Life

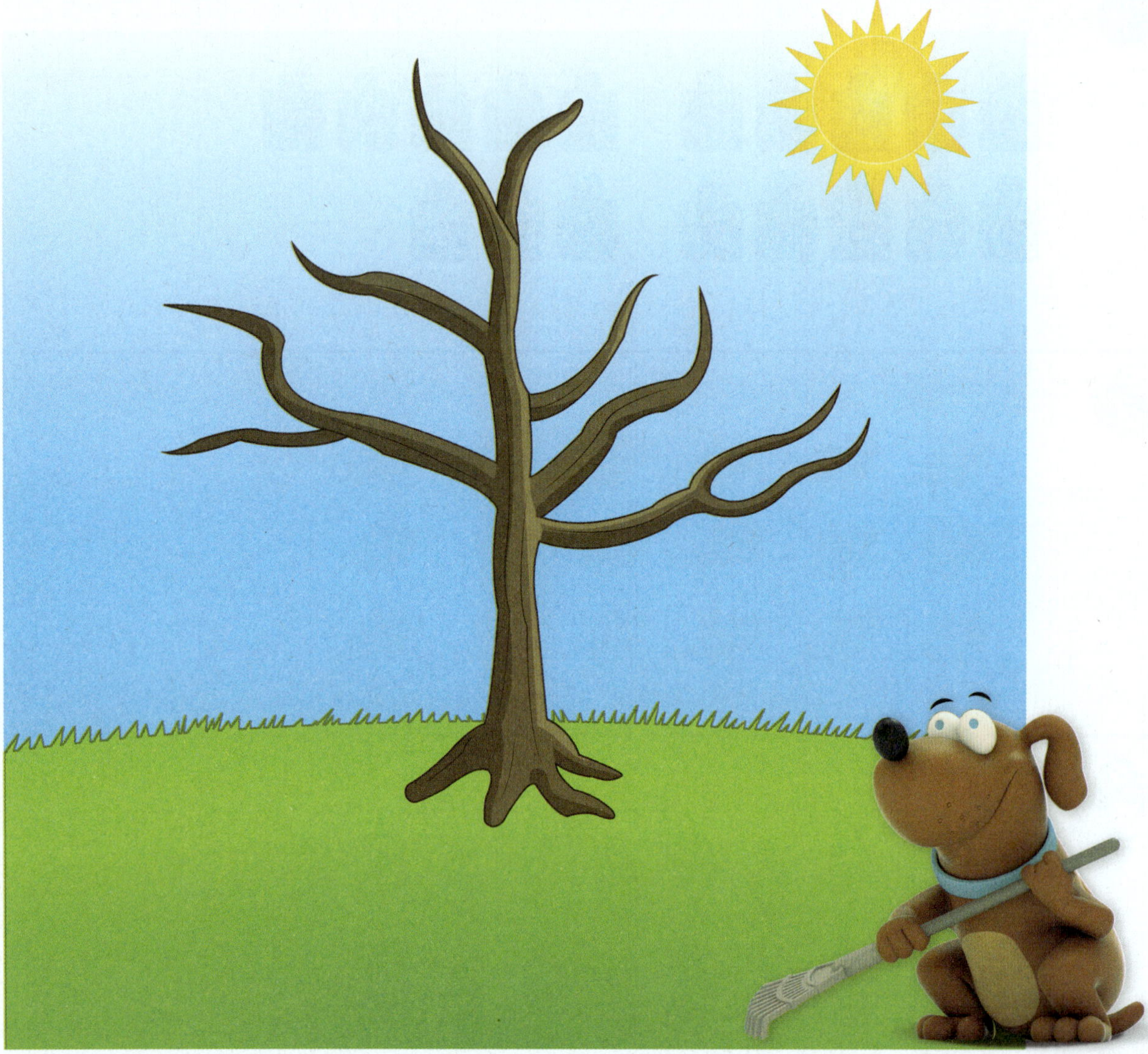

Directions: There are 19 red leaves on the ground. There are 18 orange leaves left on the tree. Draw the leaves on the ground and the leaves on the tree. Write the numbers.

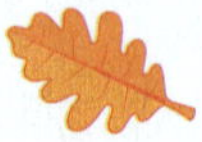

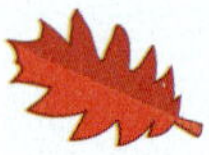

Learning Target: Count and write the numbers 18 and 19.

eighteen

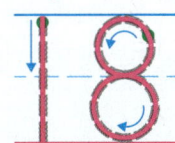

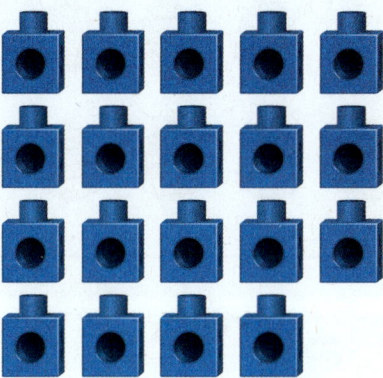

nineteen

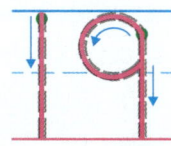

Directions: Count the objects. Say the number. Write the number.

_____

- - - - -

_____

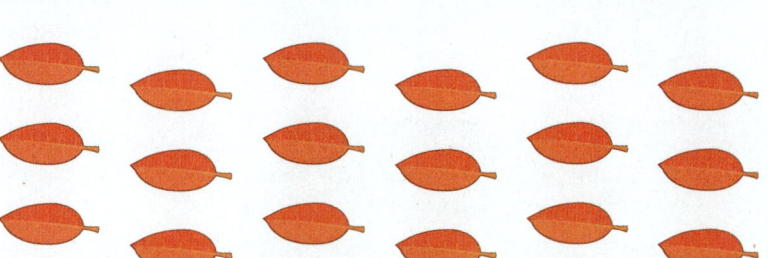

_____

- - - - -

_____

Directions: ❶ and ❷ Count the objects. Say the number. Write the number.

Directions: **3** and **4** Count the objects. Say the number. Write the number. **5** Draw 18 acorns on the ground. Write the number.

438   four hundred thirty-eight

Learning Target: Understand the numbers 18 and 19.

 **Explore and Grow**

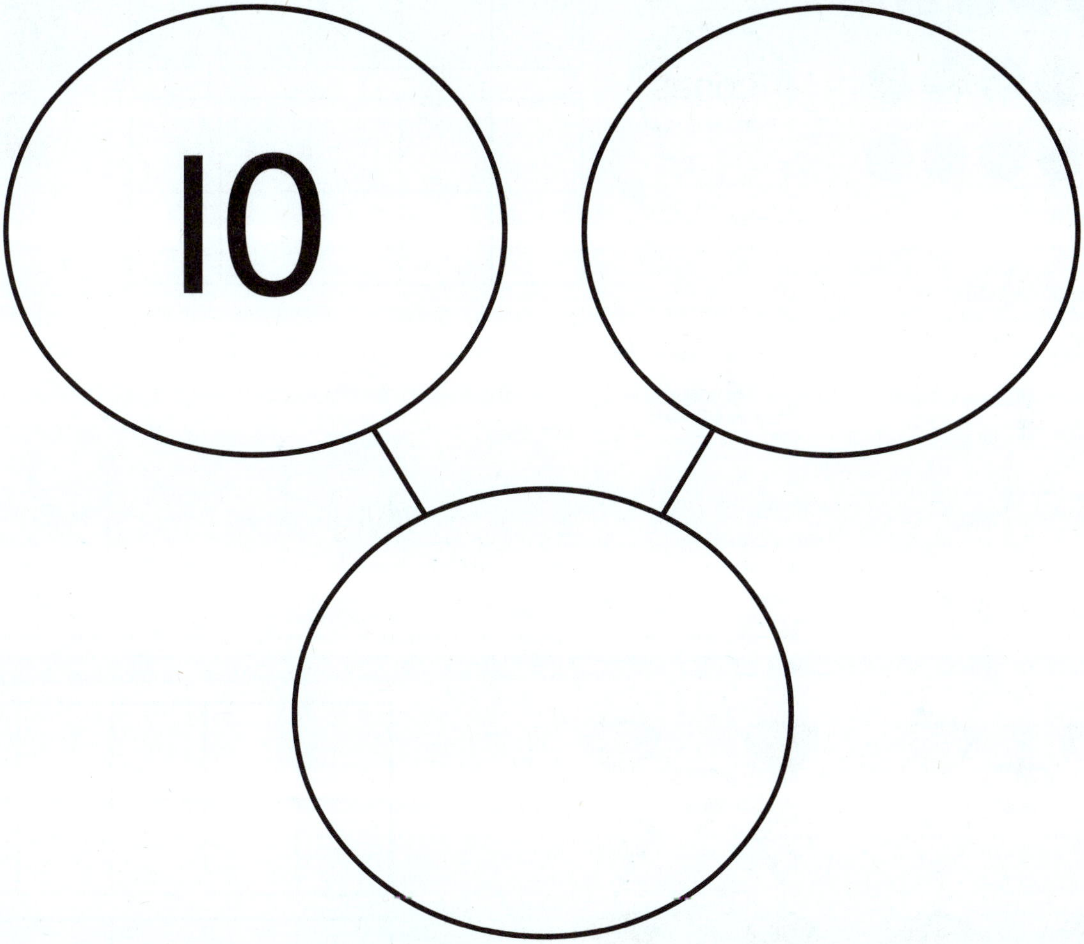

10

_10_ and _____ is _____.

Directions: Place the 10 and 8 cards as the parts on the number bond. Slide the cards and hide the zero with the 8 card to make the whole. Write the parts and the whole.

Chapter 8 | Lesson 11

four hundred thirty-nine

439

Think and Grow

10 ones

8 ones

18 == 10 + 8

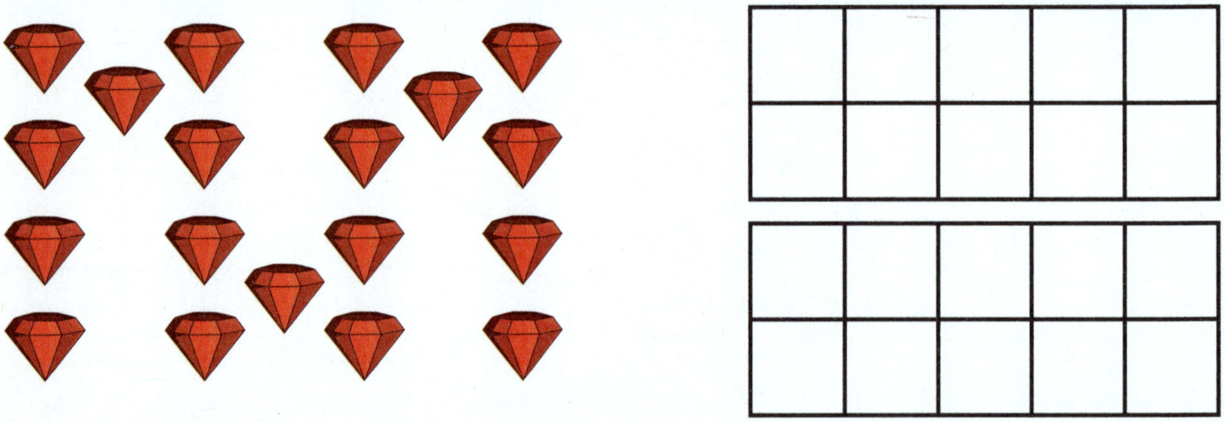

____ == 10 + ____

Directions: Circle 10 objects. Draw dots in the top ten frame to show how many objects are circled. Draw dots in the bottom ten frame to show how many more objects there are. Use the ten frames to write an addition sentence.

440   four hundred forty

✓ Apply and Grow: Practice

1

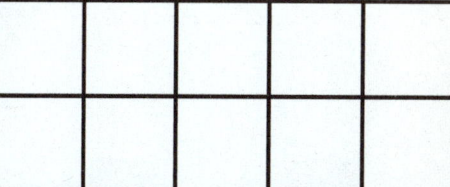

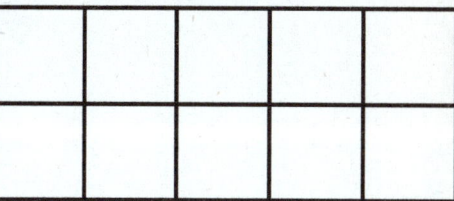

 $=$  $+$ _____

2

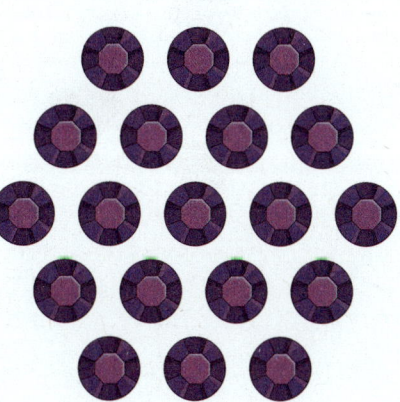

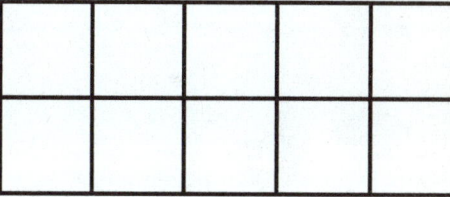

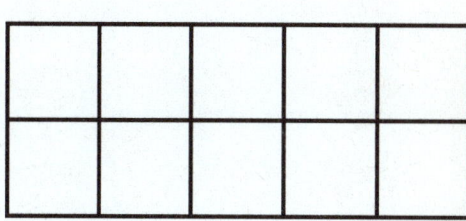

_____ $=$  $+$ _____

Directions: **1** and **2** Circle 10 gemstones. Draw dots in the top ten frame to show how many gemstones are circled. Draw dots in the bottom ten frame to show how many more gemstones there are. Use the ten frames to write an addition sentence.

Chapter 8 | Lesson 11

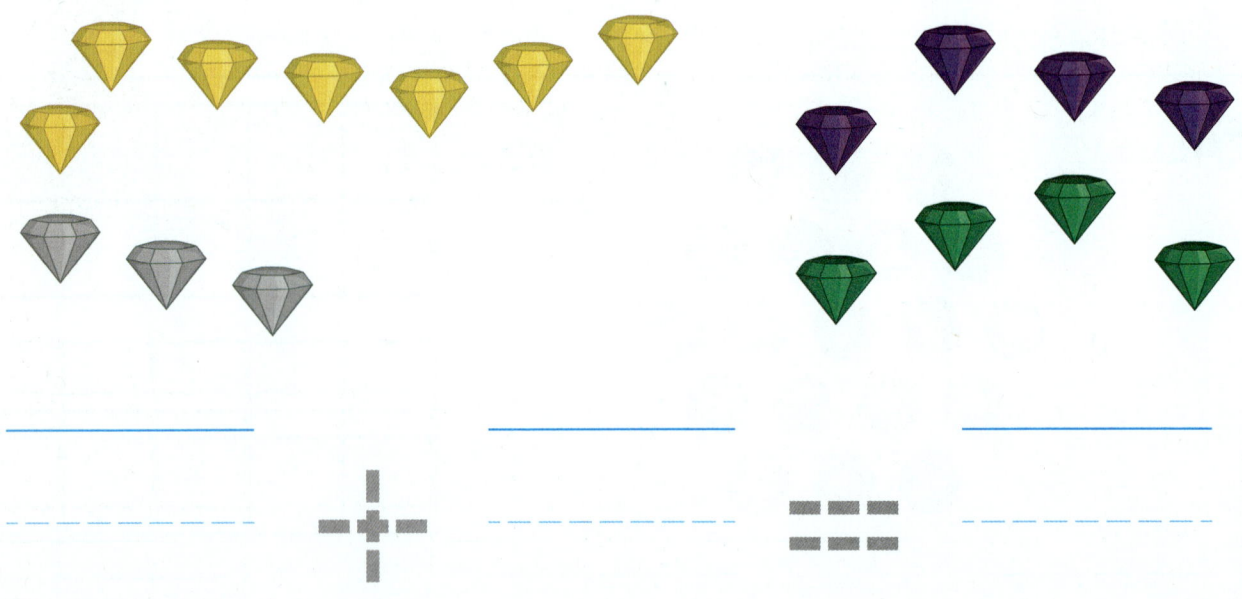

Directions:

- You have pink gemstones and blue gemstones. Your friend has silver gemstones and gold gemstones. Circle your gemstones. Write an addition sentence to match the picture. How many gemstones do you have? Circle the number.

- You have gold gemstones and silver gemstones. Your friend has purple gemstones and green gemstones. Circle your gemstones. Write an addition sentence to match the picture. How many gemstones does your friend have? Circle the number.

442   four hundred forty-two

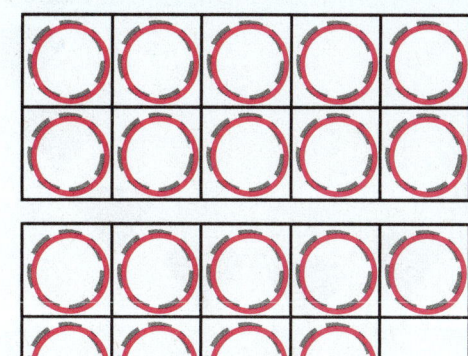

$$19 = 10 + 9$$

Directions: Circle 10 objects. Draw dots in the top ten frame to show how many objects are circled. Draw dots in the bottom ten frame to show how many more objects there are. Use the ten frames to complete the addition sentence.

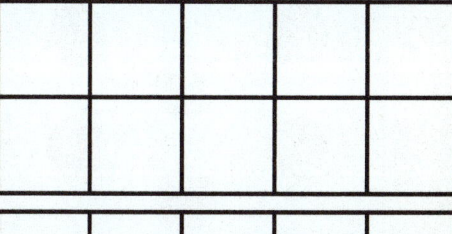

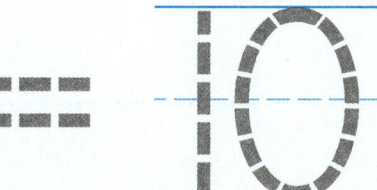

$$\_\_ = 10 + \_\_$$

Directions:  Circle 10 gemstones. Draw dots in the ten frame to show how many gemstones are circled. Draw dots in the bottom ten frame to show how many more gemstones there are. Use the ten frames to write an addition sentence.

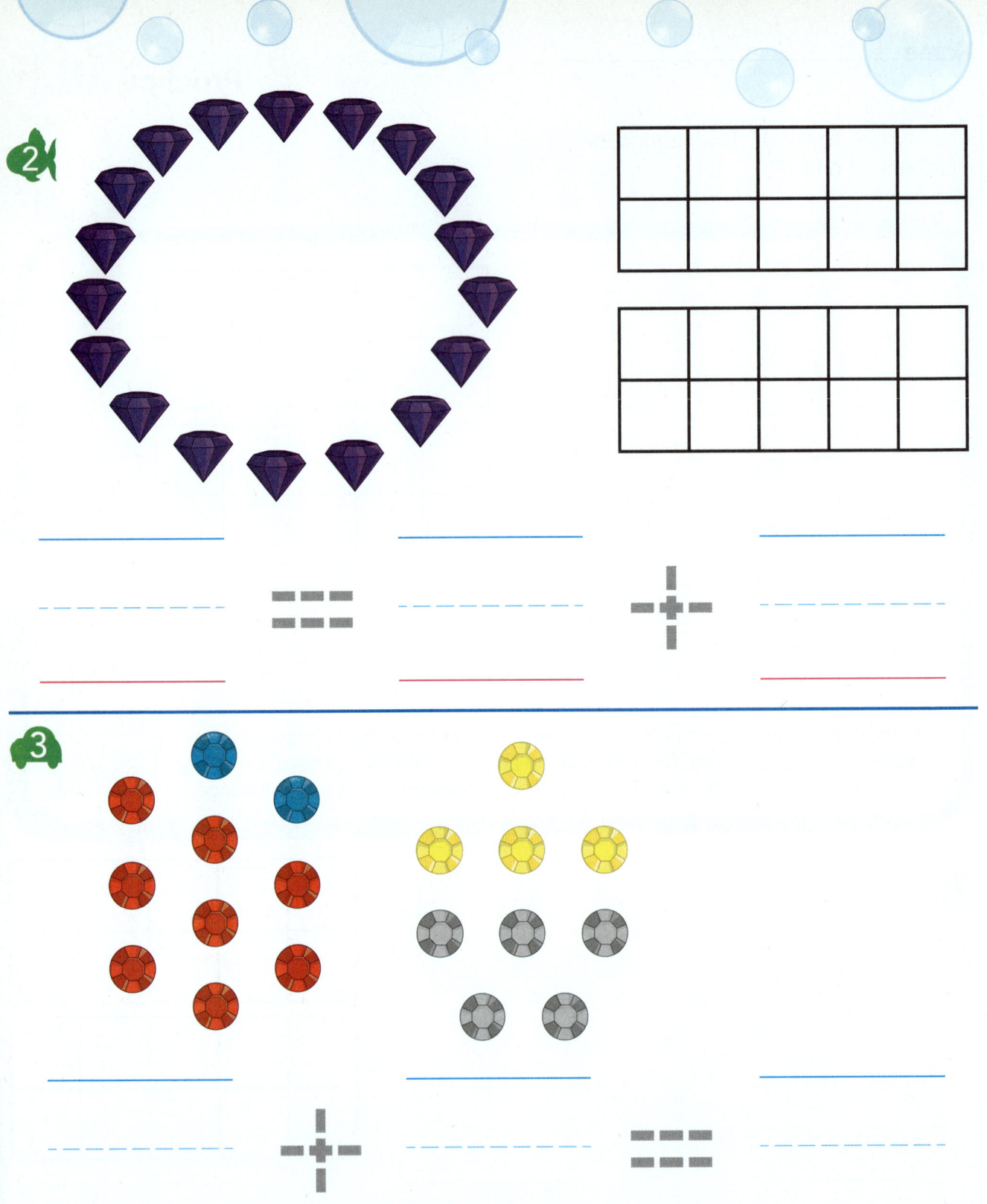

2

3

_____ ▬▬ _____ ＋ _____

_____ ＋ _____ ▬▬ _____

Directions: 🐟 Circle 10 gemstones. Draw dots in the top ten frame to show how many gemstones are circled. Draw dots in the bottom ten frame to show how many more gemstones there are. Use the ten frames to write an addition sentence.

🐢 You have red gemstones and blue gemstones. Your friend has yellow gemstones and silver gemstones. Circle your gemstones. Write an addition sentence to match the picture. How many gemstones does your friend have? Circle the number.

_____   _____   _____

- - - - - - -  - - - - - - -   = =   - - - - - - -

_____   _____   _____

_____   _____   _____

- - - - - - -  - - - - - - -   = =   - - - - - - -

_____   _____   _____

Directions: Classify the stars into 2 categories. Circle to show each group. Then write an addition sentence to tell how many stars there are in all. There are red stars and blue stars in the sky. The number of red stars is 1 more than 9. The number of blue stars is greater than 5, but less than the number of red stars. Draw and color the stars. Then write an addition sentence to tell how many stars there are in all.

Chapter 8

Number Flip and Find

Directions: Place the Number Flip and Find Cards facedown in the boxes. Take turns flipping 2 cards. If your cards show the same number, keep the cards. If your cards show different numbers, flip the cards back over. Repeat this until all cards are gone.

8.1 Identify Groups of 10

10 ones and _____ ones

 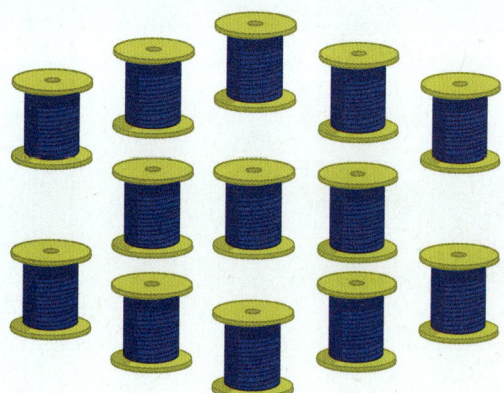

_____ ones and _____ ones

Directions: 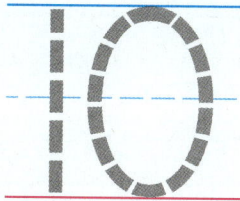 Circle 10 objects. Tell how many more objects there are. Then write the numbers.

Chapter 8

8.2 **Count and Write 11 and 12**

 3

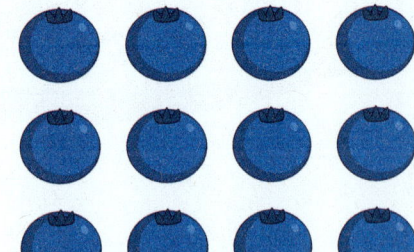

 4

 8.3 **Understand 11 and 12**

 5

 == $\boxed{10}$ + _____

Directions: **3** and **4** Count the fruit. Say the number. Write the number.
5 Circle 10 buses. Draw dots in the ten frame to show how many buses are circled. Draw dots in the five frame to show how many more buses there are. Use the frames to write an addition sentence.

8.4 Count and Write 13 and 14

6

_____

7

_____

8.5 Understand 13 and 14

8

 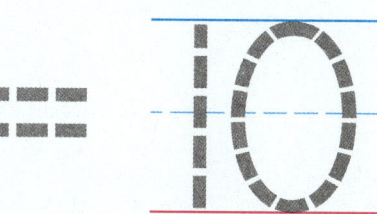 

Directions: **6** and **7** Count the vegetables. Say the number. Write the number.
8 Circle 10 hats. Draw dots in the ten frame to show how many hats are circled.
Draw dots in the five frame to show how many more hats there are. Use the frames
to write an addition sentence.

Chapter 8

8.6 Count and Write 15

8.7 Understand 15

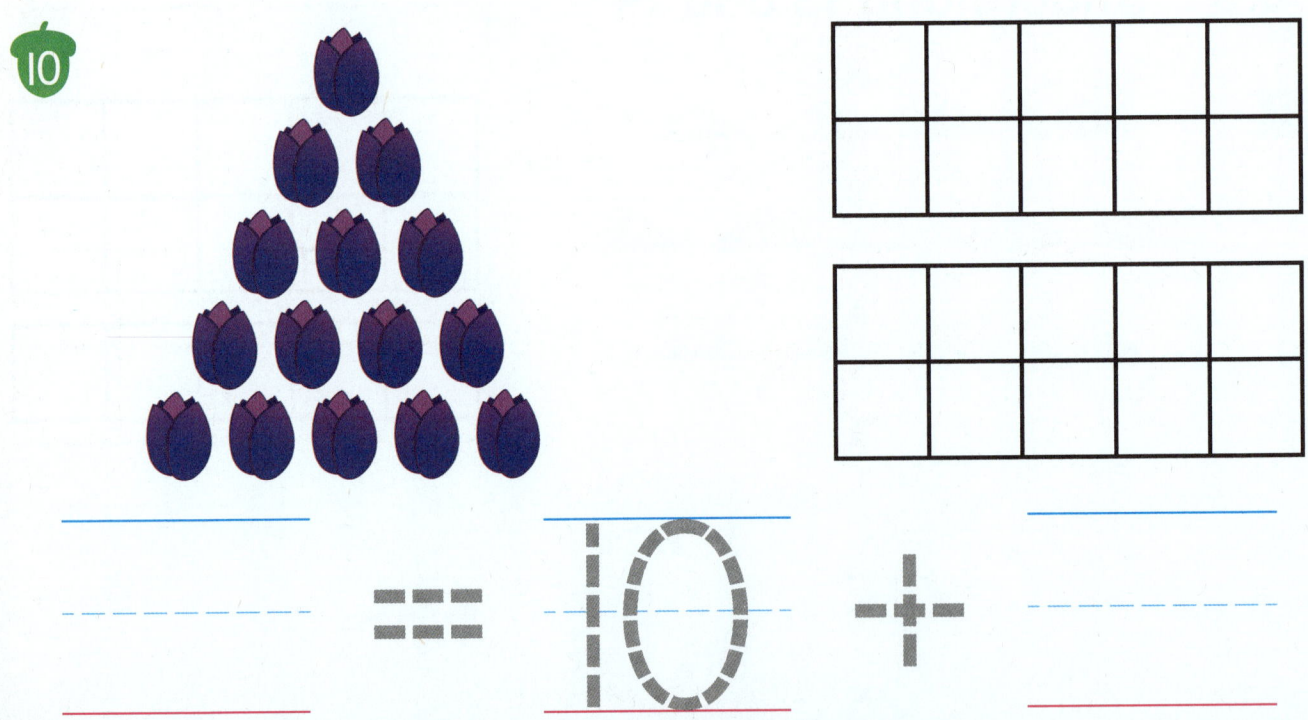

Directions: 🦆 Count the fish. Say the number. Write the number. 🌰 Circle 10 flowers. Draw dots in the top ten frame to show how many flowers are circled. Draw dots in the bottom ten frame to show how many more flowers there are. Use the ten frames to write an addition sentence.

450   four hundred fifty

Count and Write 16 and 17

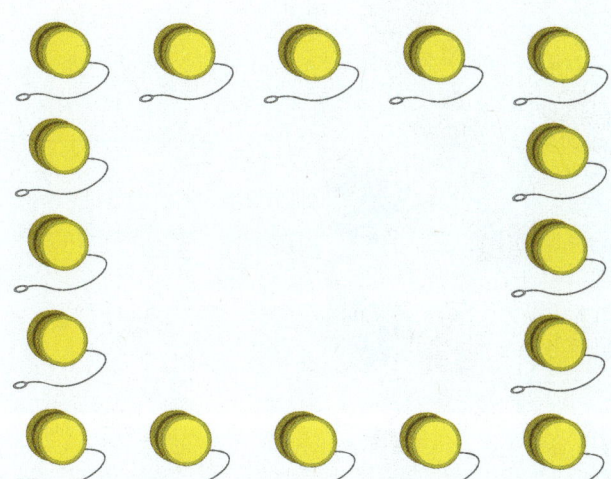

_____

- - - - - - - - - - -

_____

8.9 Understand 16 and 17

_____ = |0 + _____

Directions: Count the yo-yos. Say the number. Write the number. Circle 10 stickers. Draw dots in the top ten frame to show how many stickers are circled. Draw dots in the bottom ten frame to show how many more stickers there are. Use the ten frames to write an addition sentence.

8.10 Count and Write 18 and 19

_____

- - - - - - - - - - -

_____

8.11 Understand 18 and 19

_____   +   _____   =   _____

- -

_____   _____   _____

Directions:  Count the leaves. Say the number. Write the number. 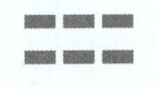 You have blue gemstones and orange gemstones. Your friend has purple gemstones and green gemstones. Circle your gemstones. Write an addition sentence to match the picture. How many gemstones do you and your friend have in all? Circle the number.

452   four hundred fifty-two

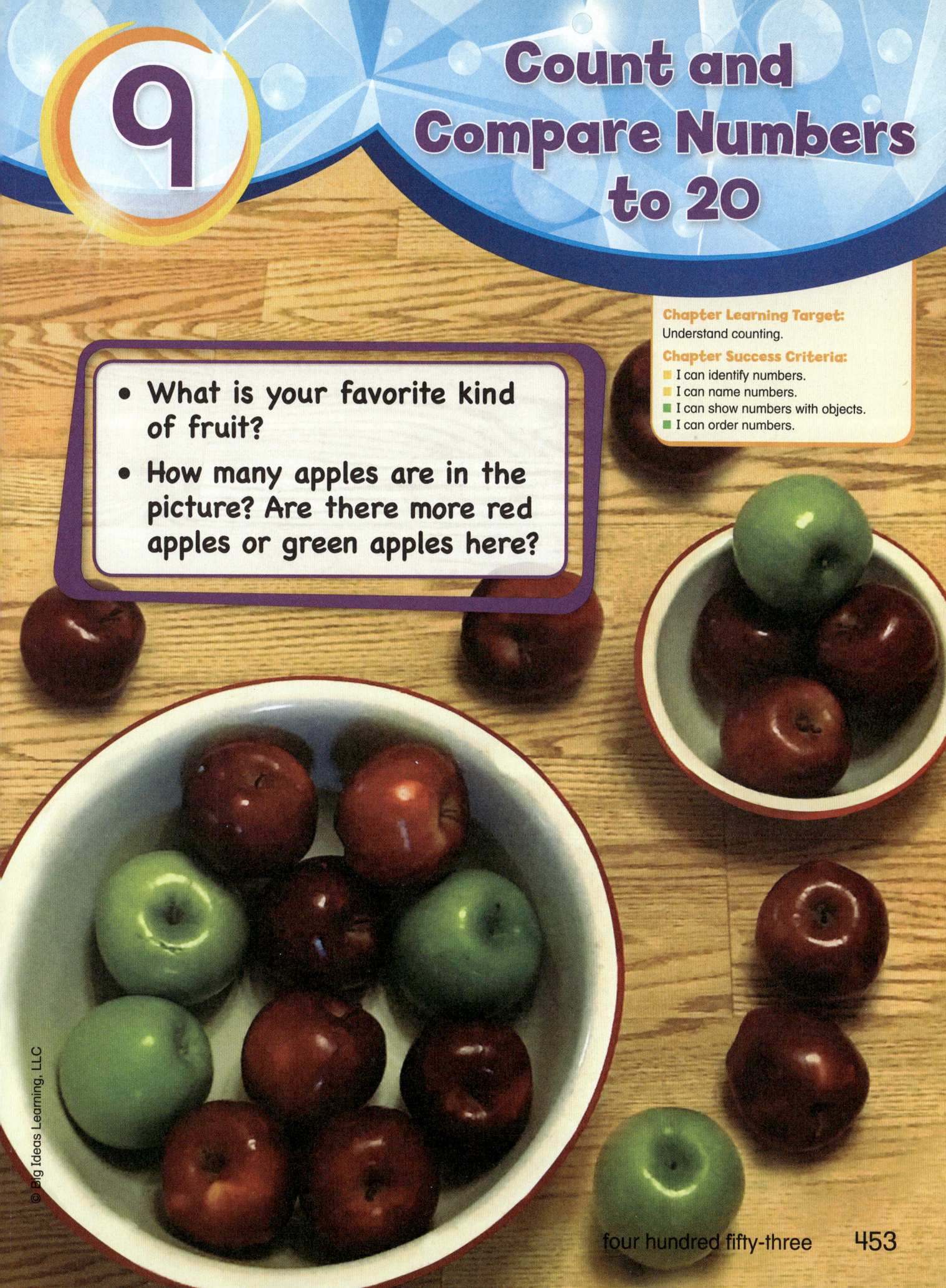

Chapter Learning Target:
Understand counting.

Chapter Success Criteria:
- I can identify numbers.
- I can name numbers.
- I can show numbers with objects.
- I can order numbers.

- What is your favorite kind of fruit?

- How many apples are in the picture? Are there more red apples or green apples here?

Vocabulary

Review Words

equal
greater than

Directions: Count the fruit in each group. Write each number. Is the number of apples equal to the number of bananas? Circle the thumbs up for *yes* or the thumbs down for *no*. Circle the number that is greater than the other number.

twenty

20

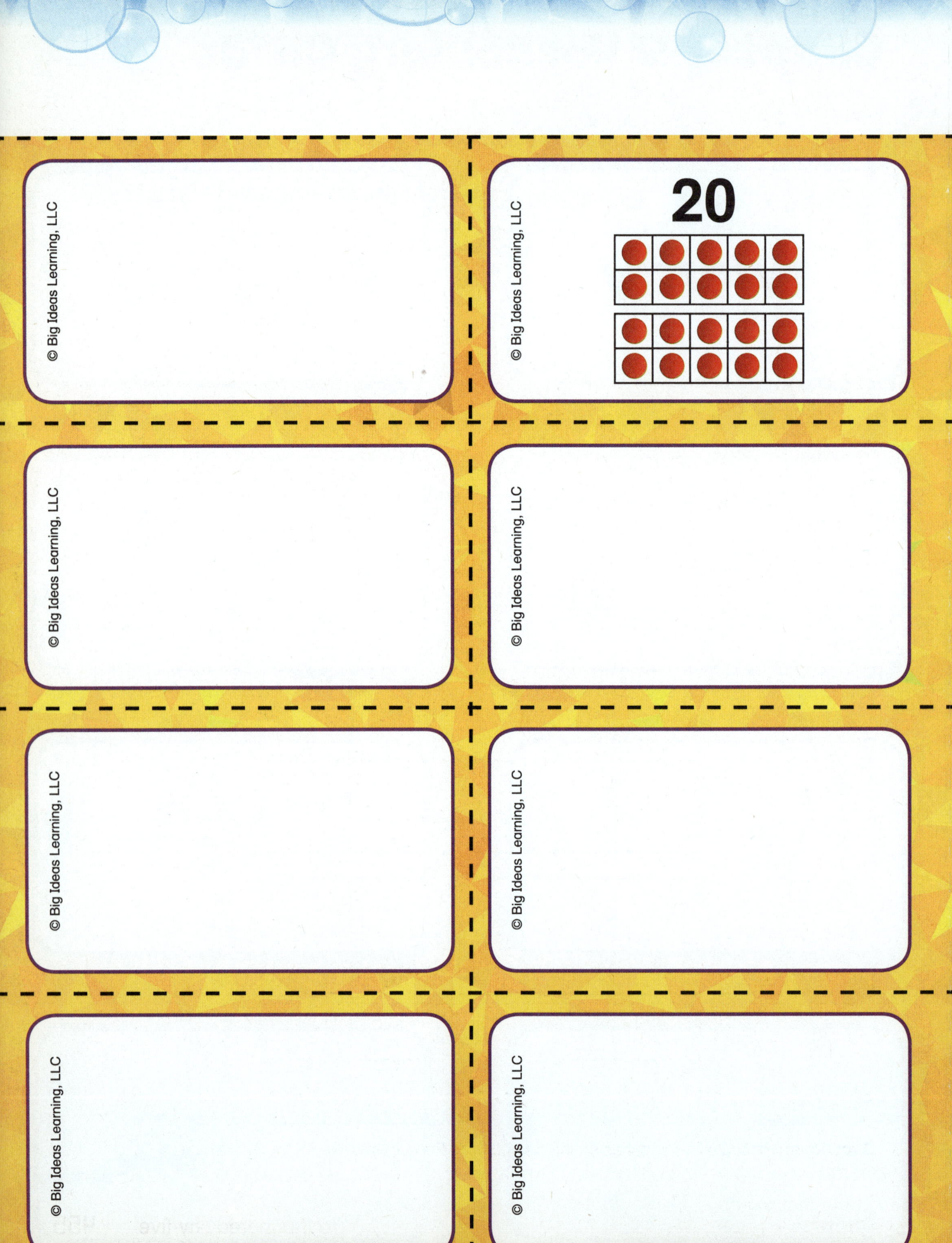

Name _____

Learning Target: Show and count the number 20.

Explore and Grow

Directions: Place 20 linking cubes on the carpet. Slide the cubes to the ten frames.

© Big Ideas Learning, LLC

Chapter 9 | Lesson 1

four hundred fifty-five 455

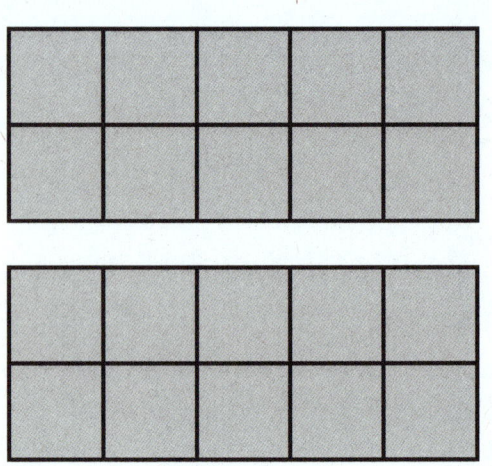

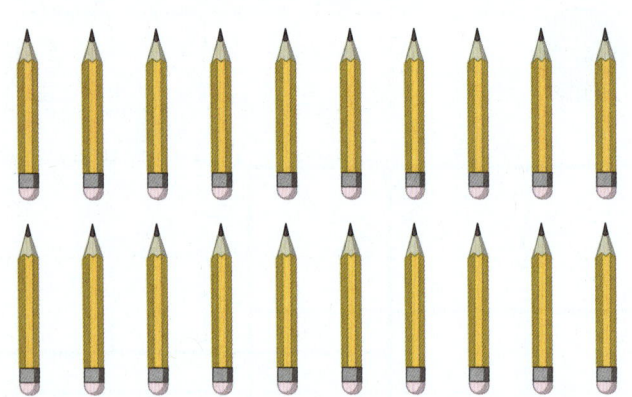

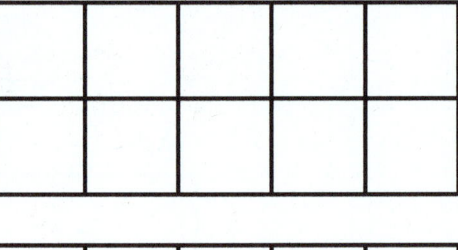

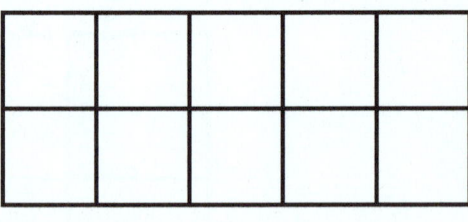

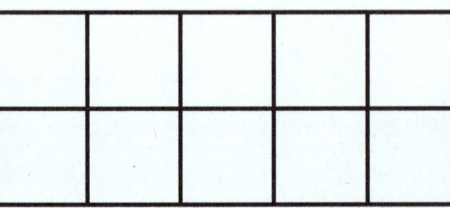

Directions: Count the objects. Color the boxes to show how many.

✓ Apply and Grow: Practice

1

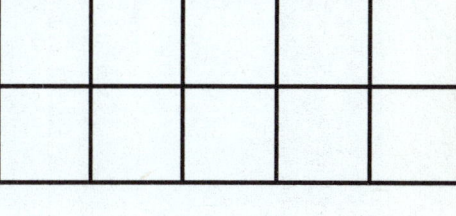

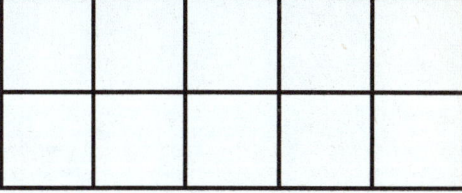

2

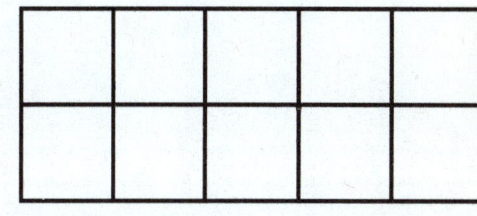

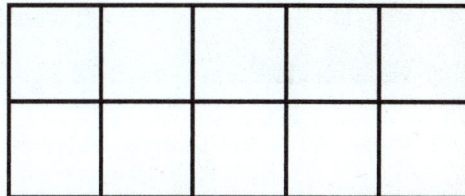

3

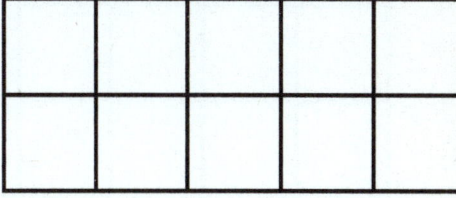

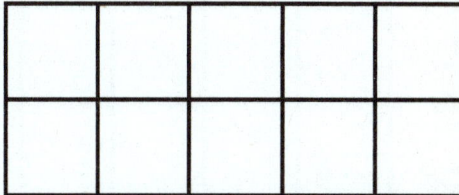

Directions: ❶–❸ Count the objects. Color the boxes to show how many.

Chapter 9 | Lesson 1

Think and Grow: Modeling Real Life

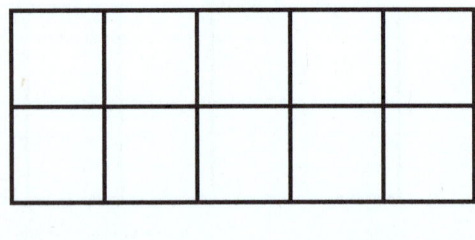

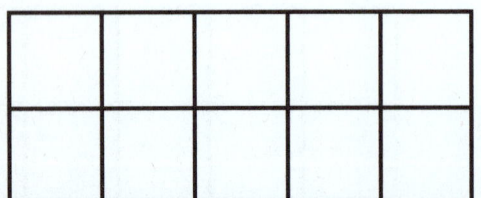

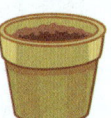

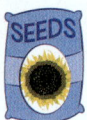

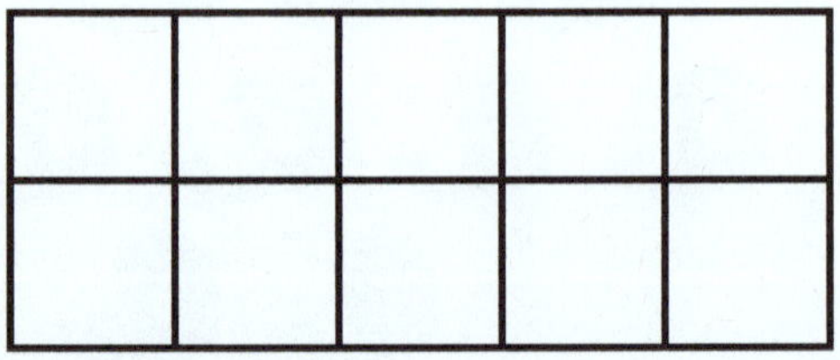

Directions: Count the objects in the picture. Color the boxes to show how many.

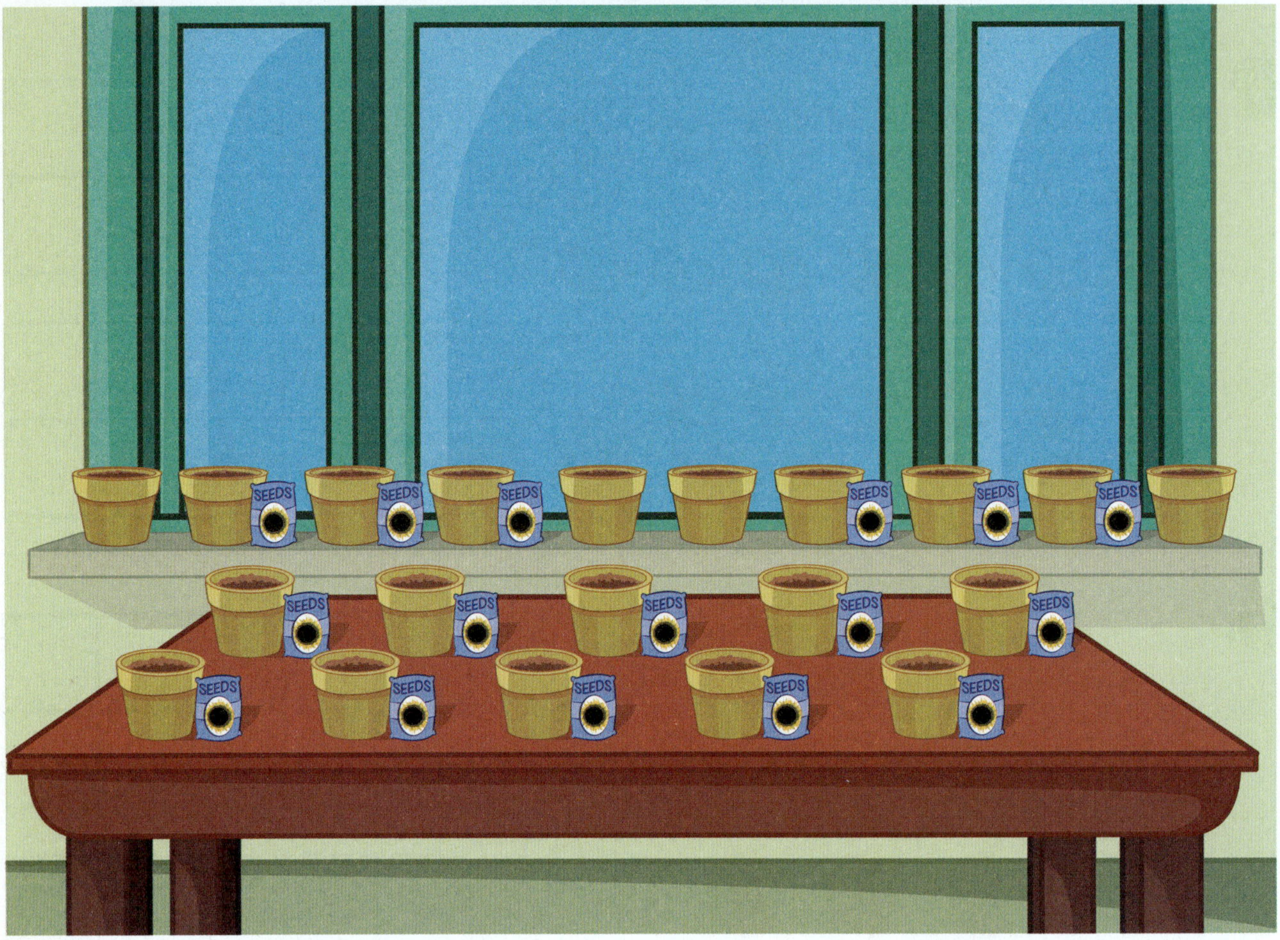

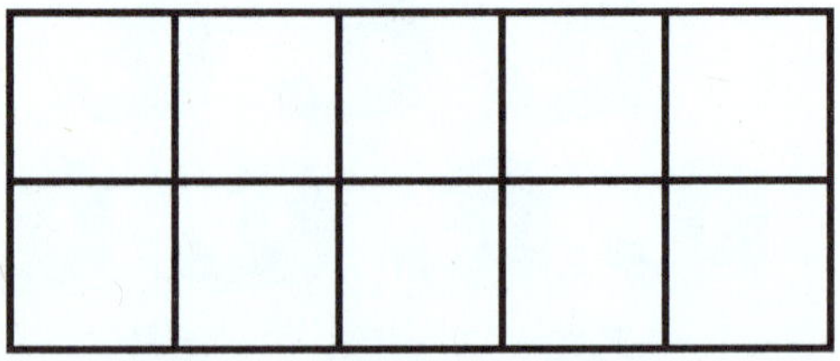

Name _____

Learning Target: Show and count the number 20.

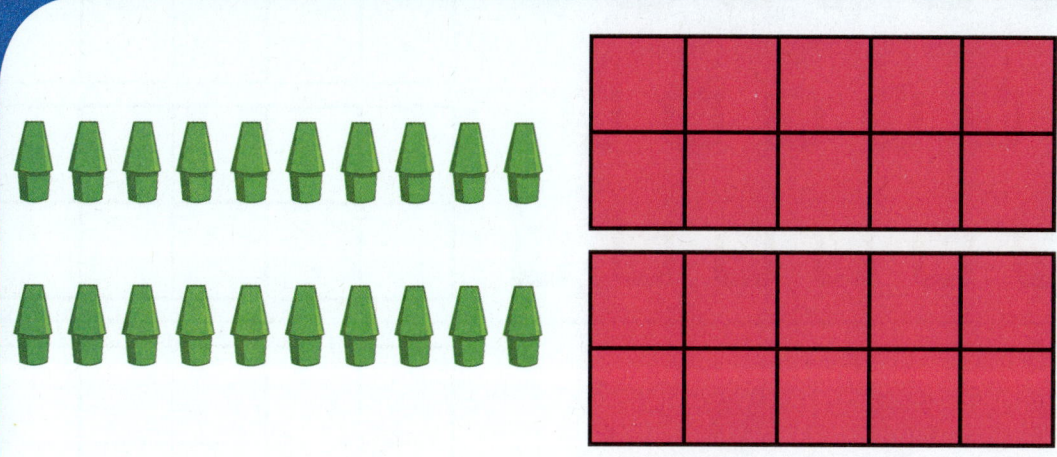

Directions: Count the erasers. Color the boxes to show how many.

1

2

Directions: **1** and **2** Count the objects. Color the boxes to show how many.

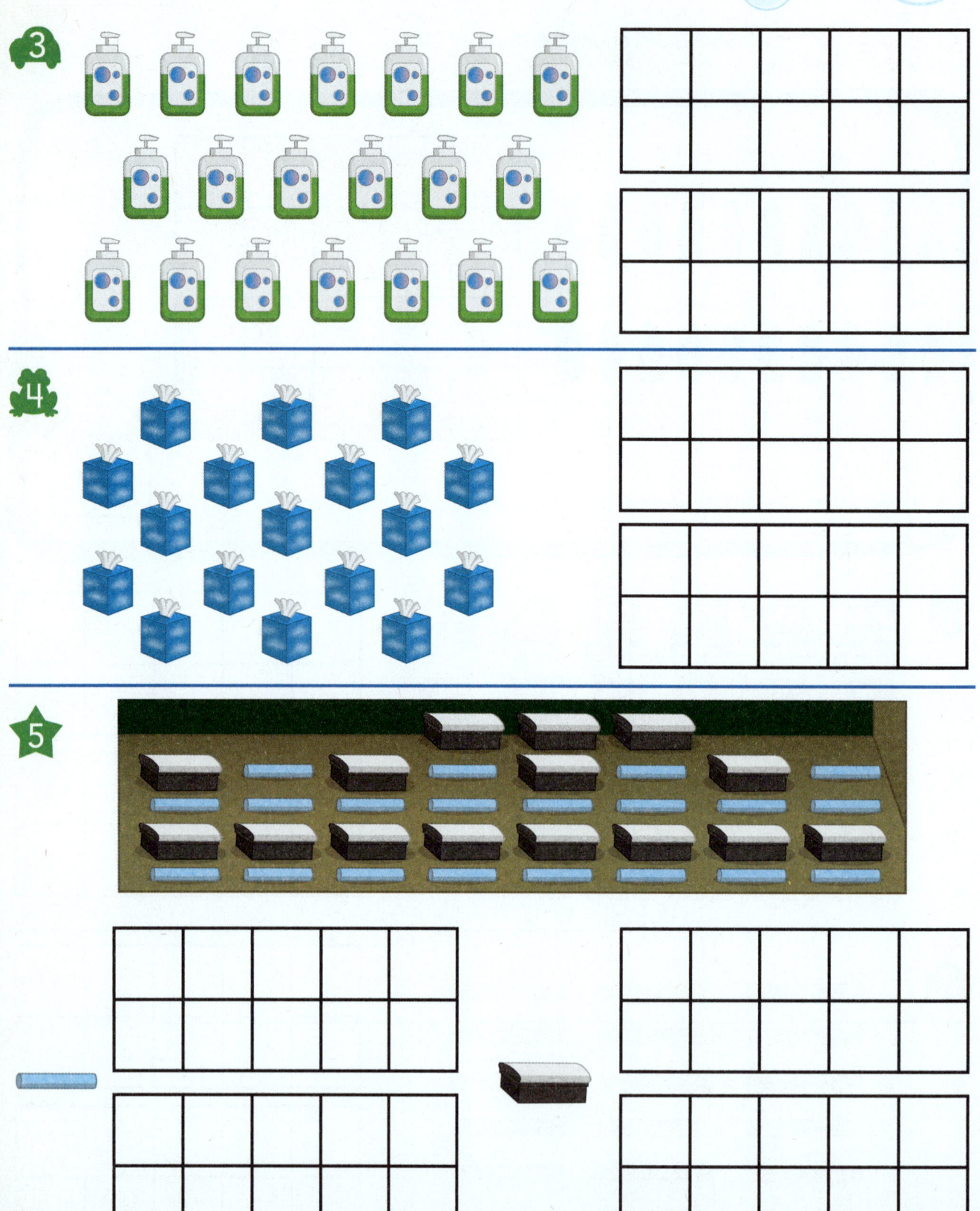

Directions: 🐢3 and 🐸4 Count the objects. Color the boxes to show how many.
⭐5 Count the objects in the picture. Color the boxes to show how many.

> **Learning Target:** Count and write the number 20.

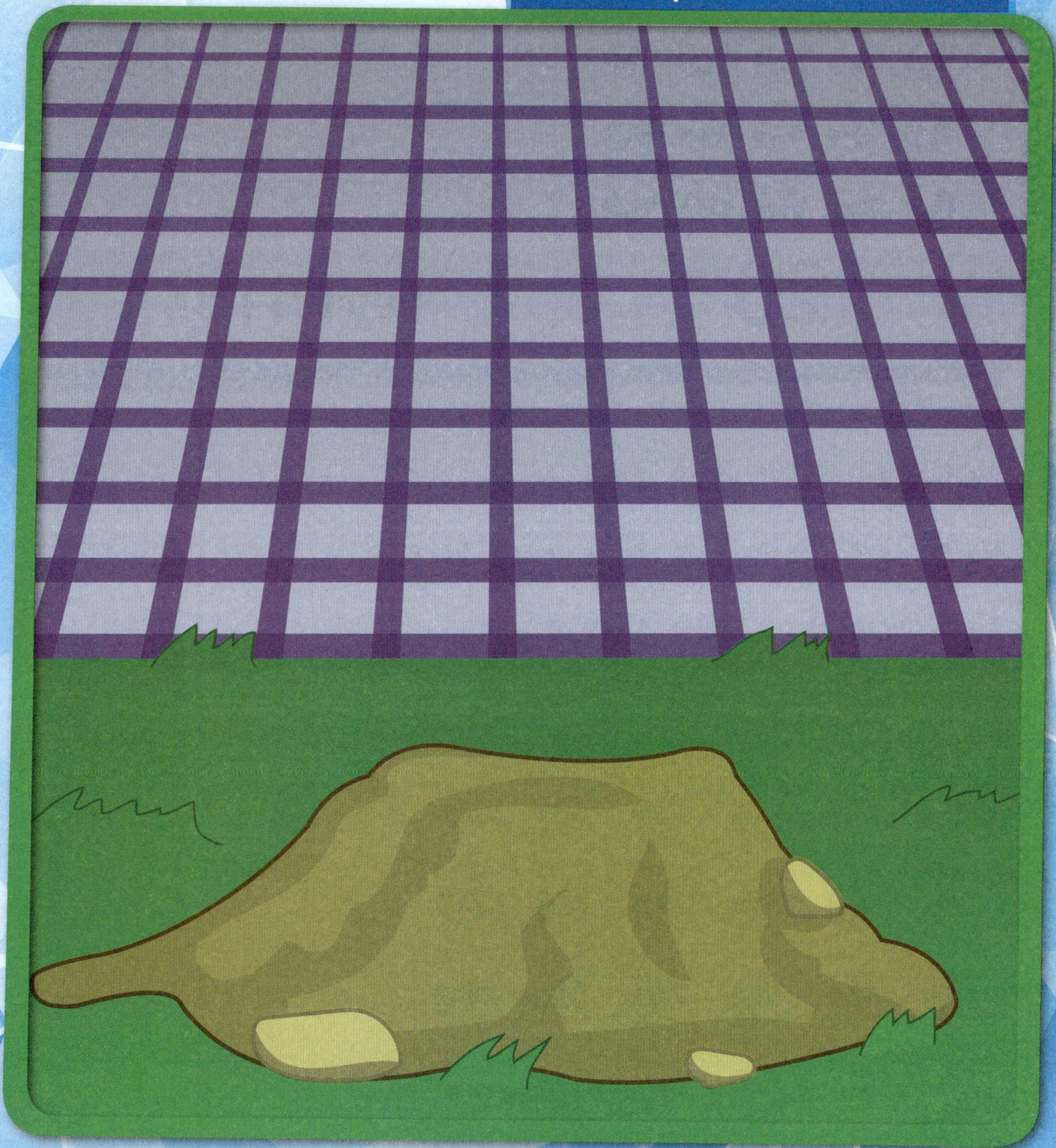

Directions: Use linking cubes to show how many ants are in the story *Ants at the Picnic.* Write how many ants are in the story.

© Big Ideas Learning, LLC

20

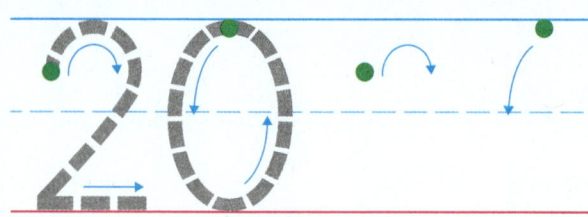

twenty

_____

_ _ _ _ _ _ _ _ _ _

_____

_____

_ _ _ _ _ _ _ _ _ _

_____

Directions:
- Count the grasshoppers. Say the number. Trace and write the number.
- Count the insects. Say the number. Write the number.

✓ Apply and Grow: Practice

 **1**

 **2**

  **3**

Directions: **1**–**3** Count the insects. Say the number. Write the number.

Think and Grow: Modeling Real Life

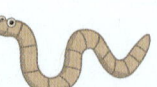

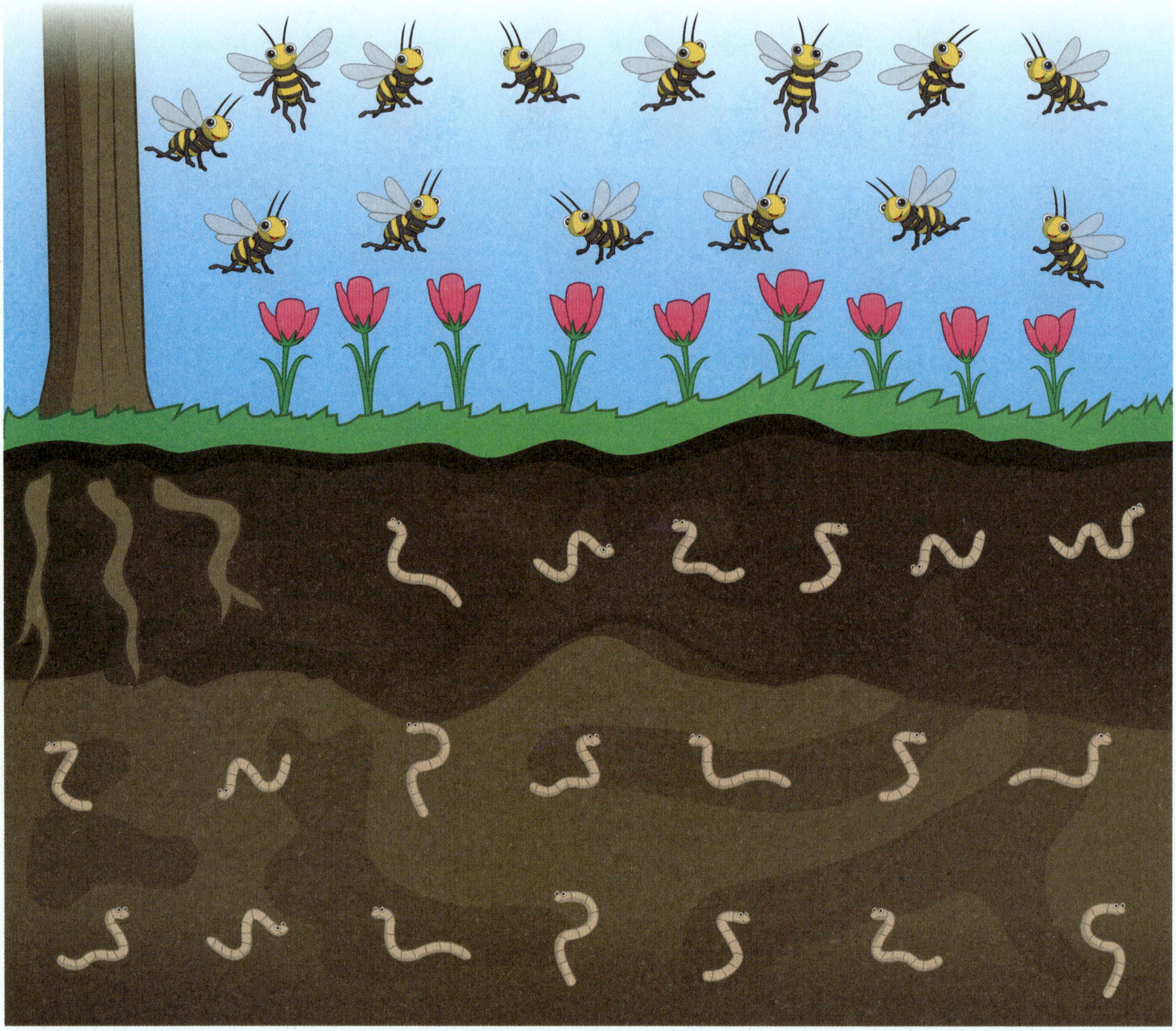

Directions: Count the animals in the picture. Say the number. Write the number.

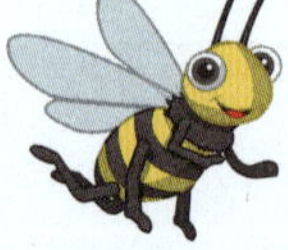

Learning Target: Count and write the number 20.

twenty

Directions: Count the insects. Say the number. Write the number.

1

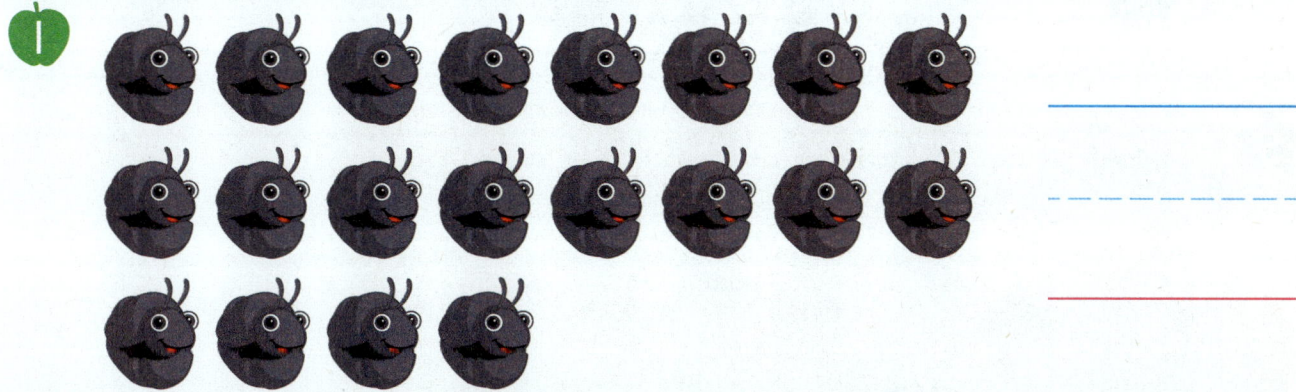

_____

- - - - - - - - - -

_____

2

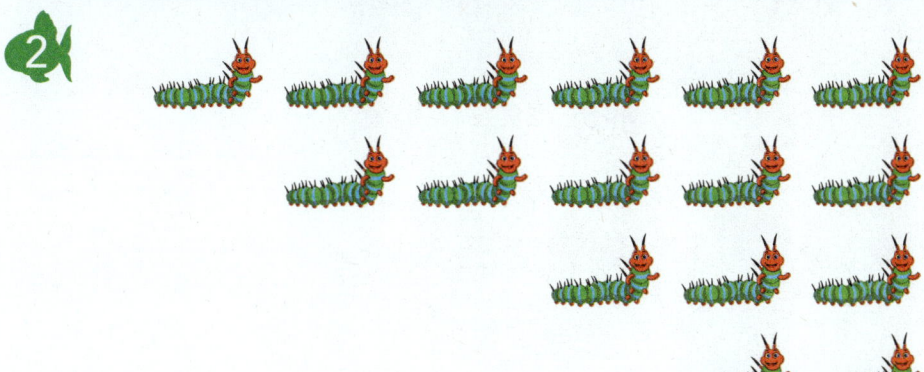

_____

- - - - - - - - - -

_____

Directions: **1** and **2** Count the insects. Say the number. Write the number.

3

_____

4

_____

5

_____

_____

Directions: **3** and **4** Count the insects. Say the number. Write the number.
5 Count the insects in the picture. Say the number. Write the number.

Name ______________________________

Count to Find How Many 9.3

Learning Target: When told a number, count that many objects.

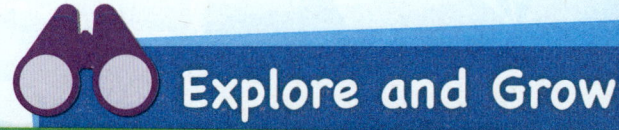

Explore and Grow

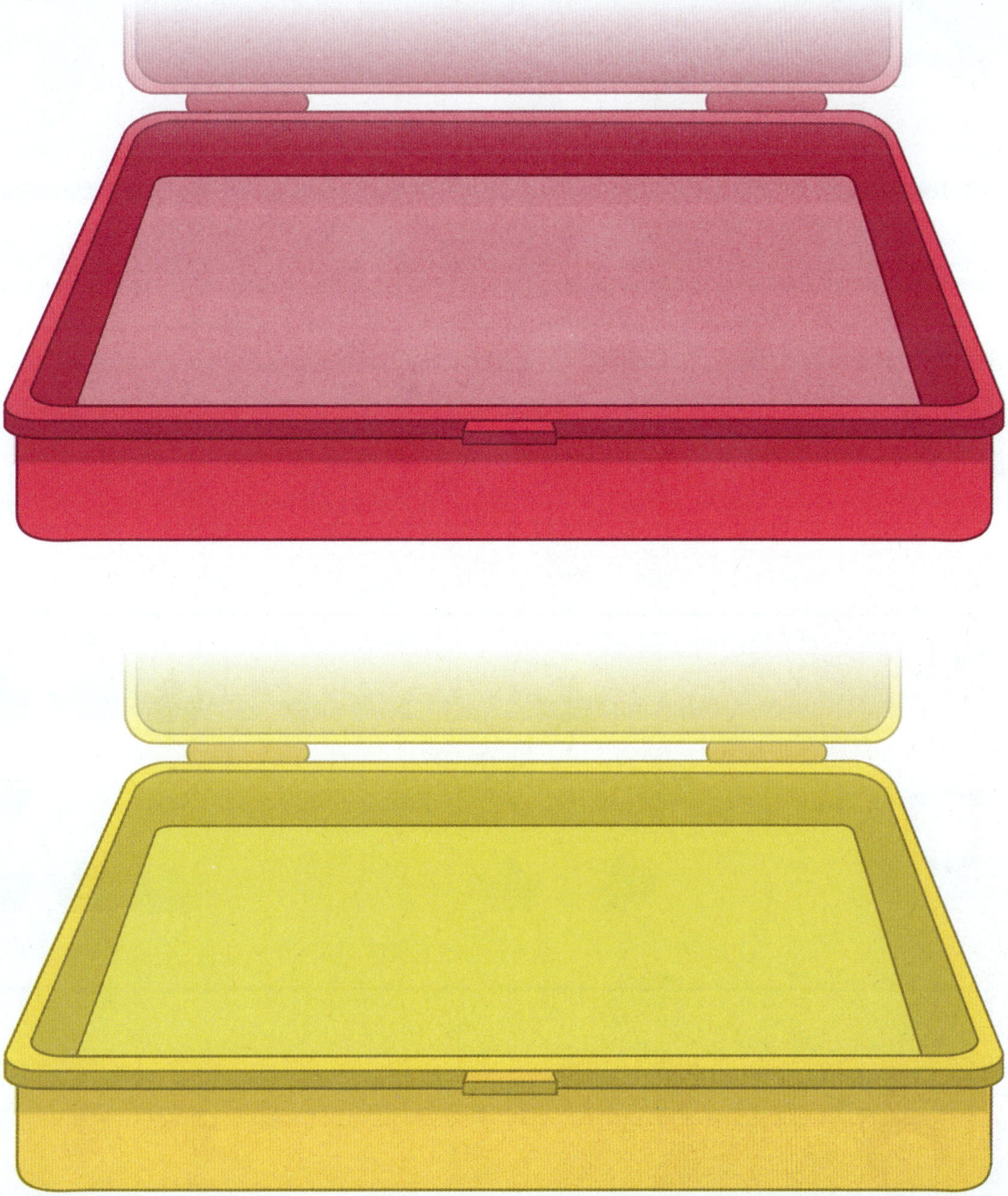

Directions: You have 14 crayons in your box. Your friend has 11 crayons in his box. Use linking cubes to show the crayons in each box.

2

17

5

12

Directions: Circle the group that has the given number of objects.

✔ Apply and Grow: Practice

 **1**

10

 2

15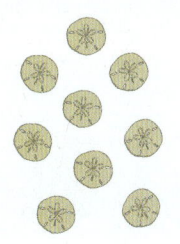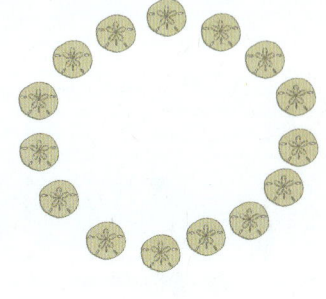

 3

19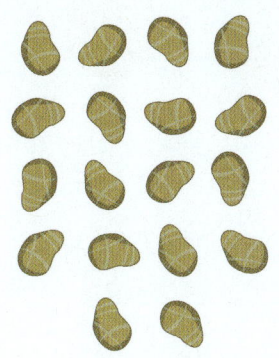

Directions: **1** – **3** Circle any group that has the given number of objects.

Think and Grow: Modeling Real Life

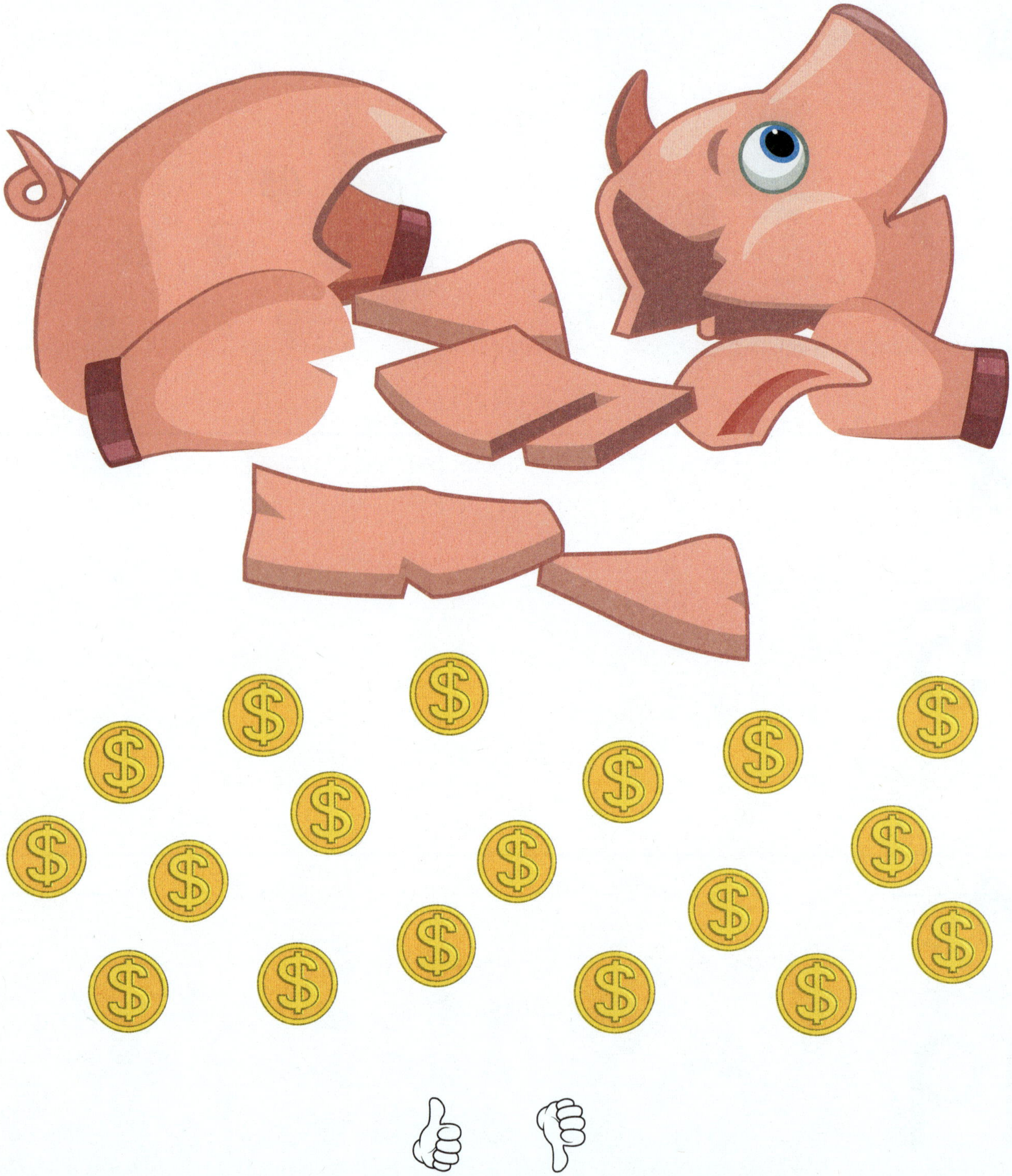

Directions: You have 20 coins in your piggy bank. You drop and break your piggy bank. Did you find all of your coins? Circle the thumbs up for *yes* or the thumbs down for *no*.

Learning Target: When told a number, count that many objects.

1

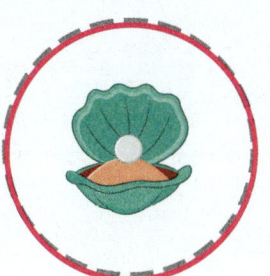  

Directions: Circle the group that has the given number of objects.

4

7

    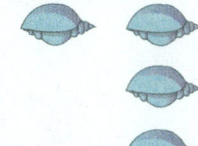

Directions: 1 and 2 Circle the group that has the given number of objects.

3

13

4

18

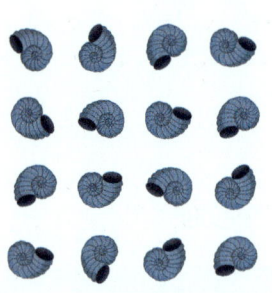  

 5

 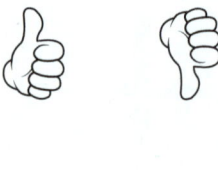

Directions: **3** and **4** Circle any group that has the given number of objects.
5 You have 20 toys in your piñata. You break your piñata. Did you find all of your toys? Circle the thumbs up for *yes* or the thumbs down for *no*.

Learning Target: Count forward from any number.

 Explore and Grow

Directions:
- Place 11 linking cubes on the ten frames. Trace the number.
- Place another cube on the ten frames. Write the number to tell how many.
- Place 1 more cube on the ten frames. Write the number to tell how many.

 **Think and Grow**

| 1 | 2 | 3 | 4 | 5 | 6 | 7 | 8 | 9 | 10 |
| 11 | 12 | 13 | 14 | 15 | 16 | 17 | 18 | 19 | 20 |

2 3 4 5 6 7

| 1 | 2 | 3 | 4 | 5 | 6 | 7 | 8 | 9 | 10 |
| 11 | 12 | 13 | 14 | 15 | 16 | 17 | 18 | 19 | 20 |

8 _____ 13

| 1 | 2 | 3 | 4 | | | | | 9 | 10 |
| 11 | 12 | 13 | 14 | 15 | 16 | 17 | 18 | 19 | 20 |

4 _____ 9

Directions: Count forward from the number in the blue circle to the number in the red circle. Write the numbers you count.

 **Apply and Grow: Practice**

1	2	3	4	5	6	7	8	⑨	
			⑭	15	16	17	18	19	20

9 _____ 14

1	2	3	4	5	⑥				
⑪	12	13	14	15	16	17	18	19	20

6 _____

3

11 _____ 16

Directions: ❶ and ❷ Count forward from the number in the blue circle to the number in the red circle. Write the numbers you count. ❸ Count forward from 11 and stop at 16. Write the numbers you count.

Chapter 9 | Lesson 4

7

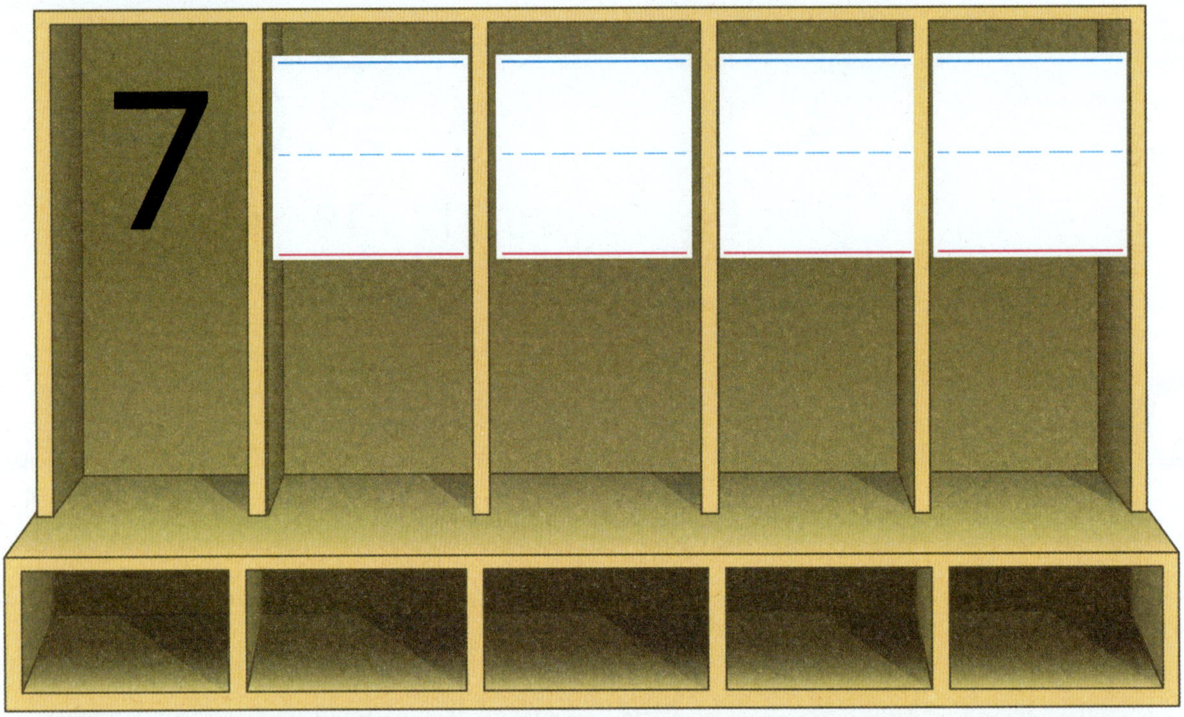

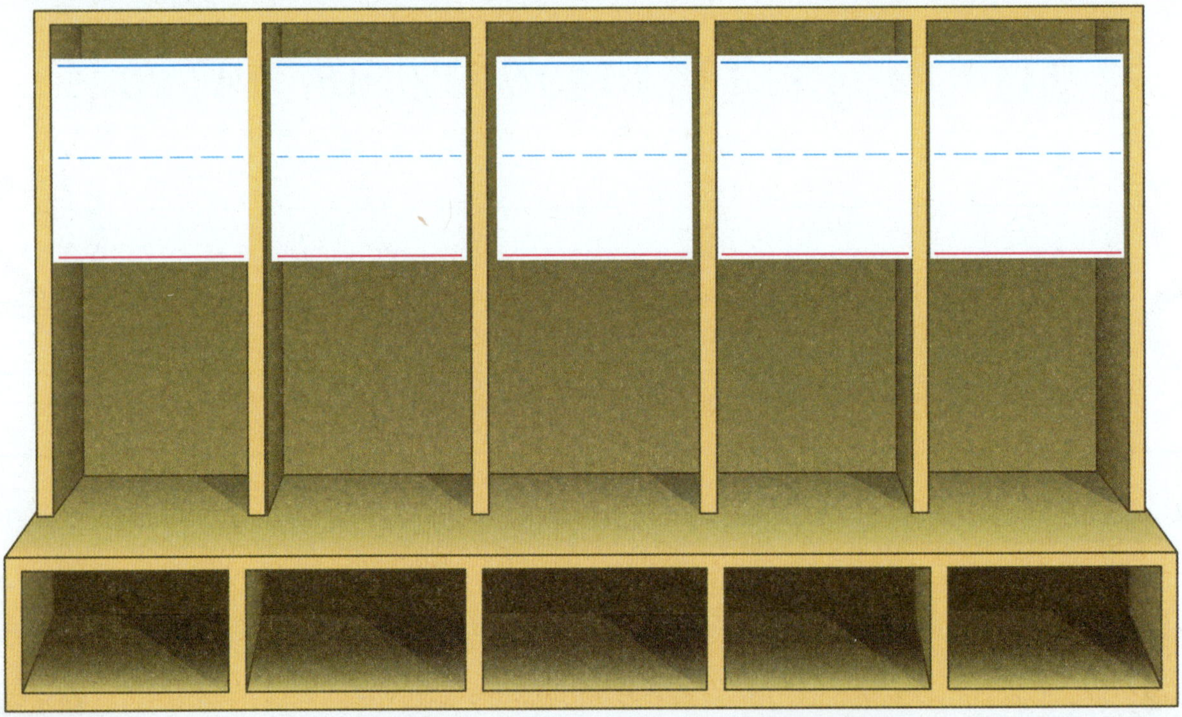

Directions:
- Your teacher labels the class cubbies and stops at 7. Count forward to finish labeling the cubbies. Write the numbers you count.
- Label the first cubby with a number from 1 to 16. Count forward from your number to finish labeling the cubbies. Write the numbers you count.

Learning Target: Count forward from any number.

①	2	3	4	5	⑥	7	8	9	10
11	12	13	14	15	16	17	18	19	20

1   2   3   4   5   6

Directions: Count forward from the number in the blue circle to the number in the red circle. Write the numbers you count.

①

1	2	3	4	5	6	⑦	8	9	10
11	⑫	13	14	15	16	17	18	19	20

7 _____ 12

 ②

1	2	③					⑧	9	10
11	12	13	14	15	16	17	18	19	20

3 _____ 8

Directions: ① and ② Count forward from the number in the blue circle to the number in the red circle. Write the numbers you count.

© Big Ideas Learning, LLC

3

1	2	3	4	5	6	7	8	9	(10)
				(15)	16	17	18	19	20

10

4

14 ___ ___ ___ ___ 19

5

12

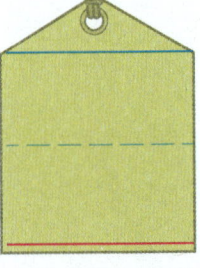

6

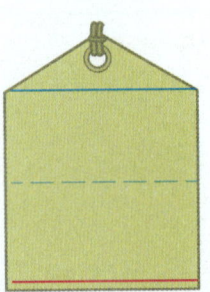

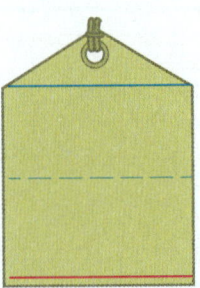

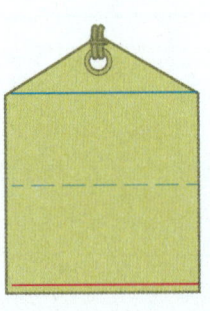

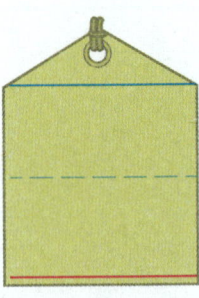

Directions: 3 Count forward from the number in the blue circle to the number in the red circle. Write the numbers you count. 4 Count forward from 14 and stop at 19. Write the numbers you count. 5 Your teacher is numbering tags and stops at 12. Count forward to finish numbering the tags. Write the numbers you count. 6 Write a number from 1 to 16 on the first tag. Count forward from your number to finish numbering the tags. Write the numbers you count.

Name _____

Learning Target: Order numbers to 20.

 **Explore and Grow**

11 1 2 13 14

16 17 ____ 19 20

Directions: Place 11 linking cubes on the ten frames. Place more cubes on the ten frames as you count forward to 20. Trace or write the missing numbers.

Directions: Count the dots in each set of ten frames. Say each number. Write each number. Then write the numbers in order.

✔ Apply and Grow: Practice

 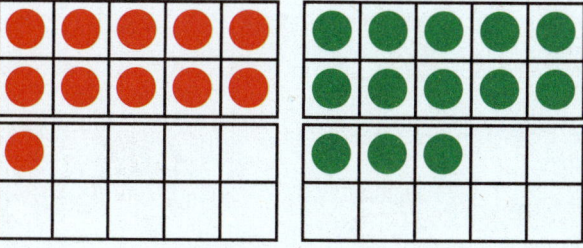 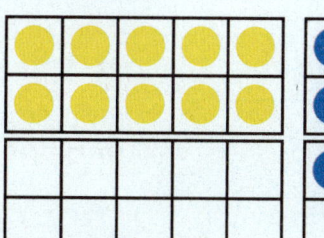 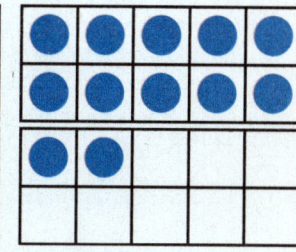

 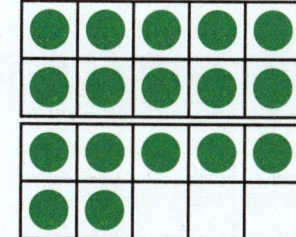 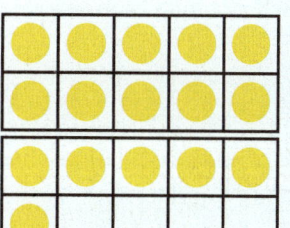 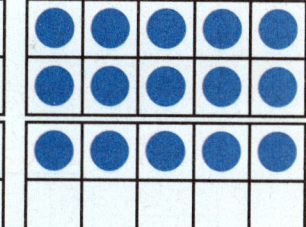

Directions: ❶ and ❷ Count the dots in each set of ten frames. Say each number. Write each number. Then write the numbers in order.

Chapter 9 | Lesson 5

four hundred eighty-one   481

Think and Grow: Modeling Real Life

Directions: Line up the students for lunch by writing the numbers in order. Circle the student who is first. Underline the student who is last.

12   14   15   13

12   13   14   15

Directions: Count the dots in each set of ten frames. Say each number. Write each number. Then write the numbers in order.

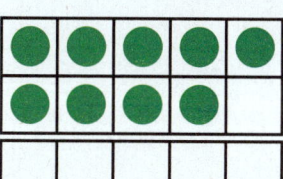
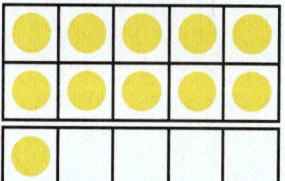
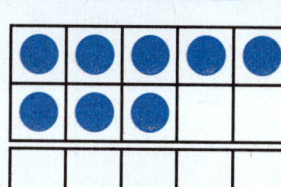

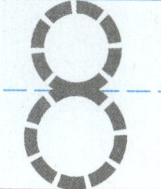

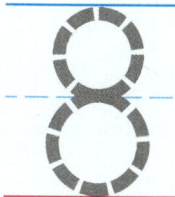
8

8

Directions: Count the dots in each set of ten frames. Say each number. Write each number. Then write the numbers in order.

2

_____   _____   _____   _____

- - - - - - - - - - -   - - - - - - - - - - -   - - - - - - - - - - -   - - - - - - - - - - -

_____   _____   _____   _____

_____   _____   _____   _____

- - - - - - - - - - -   - - - - - - - - - - -   - - - - - - - - - - -   - - - - - - - - - - -

_____   _____   _____   _____

3

_____   _____   _____   _____

- - - - - - - - - - -   - - - - - - - - - - -   - - - - - - - - - - -   - - - - - - - - - - -

_____   _____   _____   _____

Directions: **2** Count the dots in each set of ten frames. Say each number. Write each number. Then write the numbers in order. **3** Line up the train cars by writing the numbers in order. Circle the train car that is first. Underline the train car that is last.

Learning Target: Use counting to compare the numbers of objects in two groups.

 **Explore and Grow**

13

15

Directions: Place linking cubes on the ten frames to show the numbers. Which number is greater than the other number? Which number is less than the other number?

Think and Grow

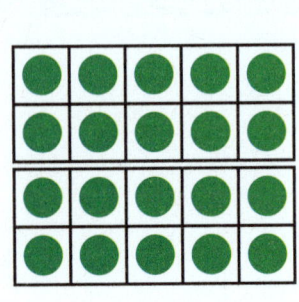 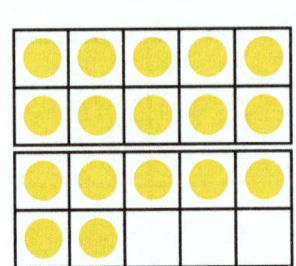 

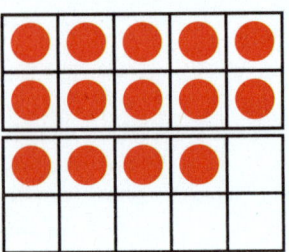 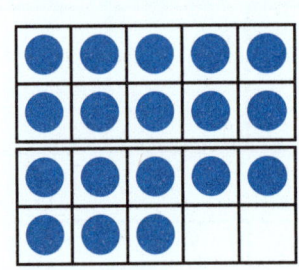 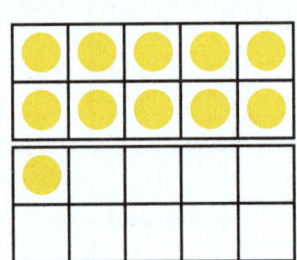

Directions: Count the dots in each set of ten frames. Write each number.
- Is the number of green dots equal to the number of yellow dots? Circle the thumbs up for *yes* or the thumbs down for *no*.
- Compare the numbers of red dots and blue dots. Circle the number that is greater than the other number.
- Compare the numbers of yellow dots and red dots. Draw a line through the number that is less than the other number.

486   four hundred eighty-six

Name _____

 **Apply and Grow: Practice**

1

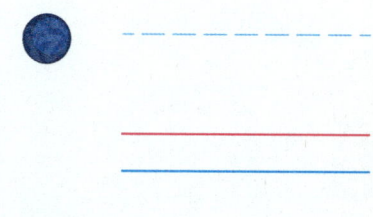

👍 👎

2

3

 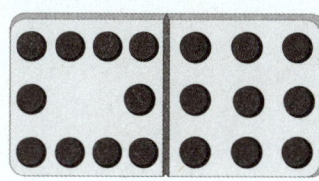

○

●

Directions: Count the objects in each group. Write each number. 🍏 Is the number of blue bouncy balls equal to the number of green bouncy balls? Circle the thumbs up for *yes* or the thumbs down for *no*. 🐟 Circle the number that is greater than the other number. 🚙 Draw a line through the number that is less than the other number.

Chapter 9 | Lesson 6

four hundred eighty-seven **487**

© Big Ideas Learning, LLC

Think and Grow: Modeling Real Life

Directions:

- You have 14 balls. Your friend has a number of balls that is 1 more than 12. Draw the balls. Write the numbers. Circle the number that is greater than the other number.

- You have 18 balls. Your friend has a number of balls that is greater than 15 and less than 17. Draw the balls. Write the numbers. Draw a line through the number that is less than the other number.

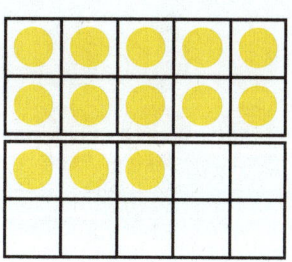

Directions: Count the dots in each set of ten frames. Write each number. Is the number of green dots equal to the number of yellow dots? Circle the thumbs up for *yes* or the thumbs down for *no*.

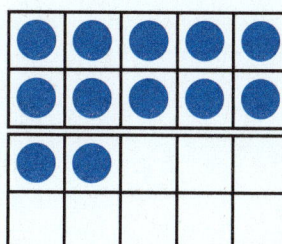 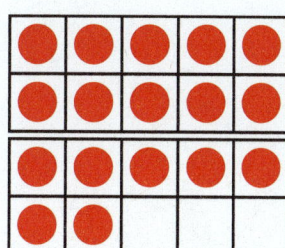

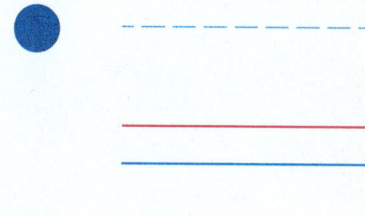

Directions: Count the dots in each set of ten frames. Write each number. Is the number of blue dots equal to the number of red dots? Circle the thumbs up for *yes* or the thumbs down for *no*.

Chapter 9 | Lesson 6

2

3

4

Who has less?

Directions: **2** Count the bouncy balls in each group. Write each number. Circle the number that is greater than the other number. **3** Count the dots on each domino. Write each number. Draw a line through the number that is less than the other number. **4** You have 13 balls. Your friend has a number of balls that is 1 more than 19. Draw the balls. Write the numbers. Draw a line through the number that is less than the other number.

Directions: Use the clues to find the number of fruit in each crate. Write each number.
- The number of apples is more than 16 but less than 18.
- The number of bananas is more than 15 but less than 17.
- The number of oranges is more than 18 but less than 20.
- The number of pears is 1 more than the number of oranges.
- The number of pineapples is 1 less than the number of oranges.
- Write the numbers in order.

Chapter 9

Number Boss

Player 1	Player 2

Directions: Each player flips a card and places it on the page. Compare the numbers. The player with the greater number takes both cards. If the numbers are equal, flip the cards again. The player with the greater number takes all the cards. Repeat until all cards have been used.

9.1 Model and Count 20

9.2 Count and Write 20

Directions: ❶ Count the paste jars. Color the boxes to show how many.
❷ Count the insects in the picture. Say the number. Write the number.

9.3 Count to Find How Many

3

11

4

9.4 Count Forward from Any Number to 20

5

1	2	3	4	5	6	7	8	9	10
11	12	13	14	15	16	17	18	19	20

5 _____ **10**

Directions: Circle any group that has the given number of objects. 🐸 You have 20 erasers in your box. You drop your box. Did you find all of your erasers? Circle the thumbs up for *yes* or the thumbs down for *no.* ⭐ Count forward from the number in the blue circle to the number in the red circle. Write the numbers you count.

494   four hundred ninety-four

6

1	2	3	4	5	6	7	8	9	10
11	12	13	14	(15)	16	17	18	19	(20)

15

9.5 Order Numbers to 20

7

Directions: **6** Count forward from the number in the blue circle to the number in the red circle. Write the numbers you count. **7** Line up the students for recess by writing the numbers in order. Circle the student who is first. Underline the student who is last.

x

 **Compare Numbers to 20**

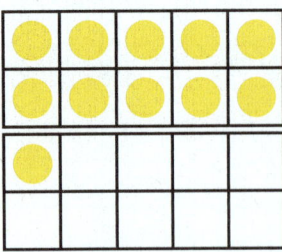 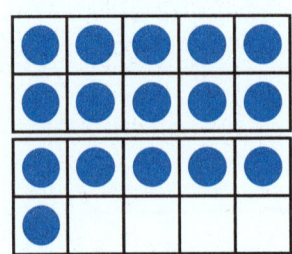

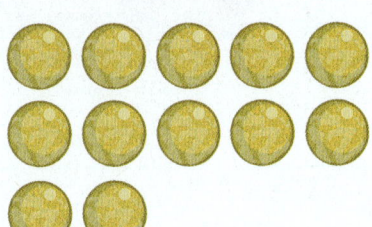

 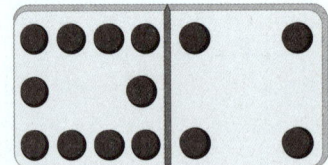

Directions: Count the objects in each group. Write each number. Is the number of yellow dots equal to the number of blue dots? Circle the thumbs up for *yes* or the thumbs down for *no*. Circle the number that is greater than the other number. Draw a line through the number that is less than the other number.

Count to 100

Chapter Learning Target:
Understand counting to 100.

Chapter Success Criteria:
- I can identify numbers.
- I can name numbers.
- I can describe numbers on a chart.
- I can explain counting numbers with patterns.

- **What events do you celebrate?**
- **How many presents are in the picture?**

Vocabulary

© Big Ideas Learning, LLC

Review Words

ten

addition sentence

_____ _____

\+

_____ _____

==

_____ _____

Directions: Circle 10 balloons. Then write an addition sentence to tell how many balloons there are in all.

Chapter 10 Vocabulary Cards

column

decade number

hundred chart

row

1	2	3	4	5	6	7	8	9	**10**
11	12	13	14	15	16	17	18	19	**20**
21	22	23	24	25	26	27	28	29	**30**
31	32	33	34	35	36	37	38	39	**40**
41	42	43	44	45	46	47	48	49	**50**
51	52	53	54	55	56	57	58	59	**60**
61	62	63	64	65	66	67	68	69	**70**
71	72	73	74	75	76	77	78	79	**80**
81	82	83	84	85	86	87	88	89	**90**
91	92	93	94	95	96	97	98	99	**100**

1	2	3	4	5	**6**	7	8	9	10
11	12	13	14	15	**16**	17	18	19	20
21	22	23	24	25	**26**	27	28	29	30
31	32	33	34	35	**36**	37	38	39	40
41	42	43	44	45	**46**	47	48	49	50
51	52	53	54	55	**56**	57	58	59	60
61	62	63	64	65	**66**	67	68	69	70
71	72	73	74	75	**76**	77	78	79	80
81	82	83	84	85	**86**	87	88	89	90
91	92	93	94	95	**96**	97	98	99	100

1	2	3	4	5	6	7	8	9	10
11	12	13	14	15	16	17	18	19	20
21	22	23	24	25	26	27	28	29	30
31	**32**	**33**	**34**	**35**	**36**	**37**	**38**	**39**	**40**
41	42	43	44	45	46	47	48	49	50
51	52	53	54	55	56	57	58	59	60
61	62	63	64	65	66	67	68	69	70
71	72	73	74	75	76	77	78	79	80
81	82	83	84	85	86	87	88	89	90
91	92	93	94	95	96	97	98	99	100

1	2	3	4	5	6	7	8	9	10
11	12	13	14	15	16	17	18	19	20
21	22	23	24	25	26	27	28	29	30
31	32	33	34	35	36	37	38	39	40
41	42	43	44	45	46	47	48	49	50
51	52	53	54	55	56	57	58	59	60
61	62	63	64	65	66	67	68	69	70
71	72	73	74	75	76	77	78	79	80
81	82	83	84	85	86	87	88	89	90
91	92	93	94	95	96	97	98	99	100

Learning Target: Count to 30 by ones.

 **Explore and Grow**

1	2	3	4	5	6	7	8	9	10
11	12	13	14	15	16	17	18	19	20
21	22	23	24	25	26	27	28	29	30

Directions: Point to each number as you count to 30. Color the number 30.

1	2	3	4	5	6	7	8	9	10
11	12	13	14	15		17	18	19	20
21	22	23	24	25	26	27	28	29	30

7

14

16

1	2	3	4	5	6	7	8	9	10
11	12	13	14	15	16	17	18	19	20
21	22		24	25	26	27	28	29	30

14

23

12

1	2	3	4	5	6	7	8		10
11	12	13	14	15	16	17	18	19	20
21	22	23	24	25	26	27	28	29	30

9

18

6

1	2	3	4	5	6	7	8	9	10
11	12	13	14	15	16	17	18	19	
21	22	23	24	25	26	27	28	29	30

11

20

29

Directions: Circle the missing number. Count to 30 starting with that number. Color the boxes as you count.

 **Apply and Grow: Practice**

1	2	3	4	5	6	7	8	9	10
11	12	13		15	16	17	18	19	20
21	22	23	24	25	26	27	28	29	30

14

15

24

1	2	3	4	5	6	7	8	9	10
11	12	13	14	15	16	17	18	19	20
21	22	23	24	25				29	30

16, 17, 18 | 26, 27, 28 | 17, 18, 19

1	2	3	4	5	6	7	8	9	10
11	12	13	14	15	16	17	18	19	20
21	22	23	24	25	26	27	28	29	30

Directions: ① Circle the missing number. Count to 30 starting with that number. Color the boxes as you count. ② Circle the missing numbers. Tell how the missing numbers are alike. ③ Find and circle the number *twenty-three*.

JUNE

Sunday	Monday	Tuesday	Wednesday	Thursday	Friday	Saturday
			1	2	🎓	4
5	6	7	8	9	10	11
12	13	🇺🇸	15	16	17	18
19	20	21	22	23	24	25
26	27	28	🎈	30		

🇺🇸 6 18 14	🎓 3 9 5	🎈 23 29 21

Directions: Circle the missing date for each sticker on the calendar. Circle the sticker that covers the earliest missing date. Underline the sticker that covers the latest missing date.

502   five hundred two

Learning Target: Count to 30 by ones.

1	2	3	4	5	6	7	8	9	10
11	12	13	14	15	16	17		19	20
21	22	23	24	25	26	27	28	29	30

(18)   27   9

Directions: Circle the missing number. Count to 30 starting with that number. Color the boxes as you count.

1

1	2	3	4	5	6	7	8	9	10
11	12	13	14	15	16	17	18	19	20
21	22	23	24	25	26	27		29	30

26

19

28

2

1	2	3	4	5	6	7	8	9	10
11	12		14	15	16	17	18	19	20
21	22	23	24	25	26	27	28	29	30

4

22

13

Directions: **1** and **2** Circle the missing number. Count to 30 starting with that number. Color the boxes as you count.

3

1	2	3	4	5	6	7	8	9	10
11	12	13	14	15	16				20
21	22	23	24	25	26	27	28	29	30

27, 28, 29 | 8, 9, 10 | 17, 18, 19

4

1	2	3	4	5	6	7	8	9	10
11	12	13	14	15	16	17	18	19	20
21	22	23	24	25	26	27	28	29	30

5

FEBRUARY

Sunday	Monday	Tuesday	Wednesday	Thursday	Friday	Saturday
8	9	10	11	12		14
15		17	18	19	20	21

 10   15   16      19   13   14

Directions: **3** Circle the missing numbers. Tell how the missing numbers are alike. **4** Find and circle the number *fourteen*. **5** Circle the missing date for each sticker on the calendar. Circle the sticker that covers the earliest missing date. Underline sticker that covers the latest missing date.

Name _____

Learning Target: Count to 50 by ones.

 **Explore and Grow**

1	2	3	4	5	6	7	8	9	10
11	12	13	14	15	16	17	18	19	20
21	22	23	24	25	26	27	28	29	30
31	32	33	34	35	36	37	38	39	40
41	42	43	44	45	46	47	48	49	50

Directions: Point to each number as you count to 50. Color the number 50.

1	2	3	4	5	6	7	8	9	10
11	12	13	14	15	16	17	18	19	20
21	22	23	24	25	26	27	28	29	30
31		33	34	35	36	37	38	39	40
41	42	43	44	45	46	47	48	49	50

23

(32)

41

1	2	3	4	5	6	7	8	9	10
11	12	13	14	15	16	17	18	19	20
21	22	23	24	25	26		28	29	30
31	32	33	34	35	36	37	38	39	40
41	42	43	44	45	46	47	48	49	50

36

18

27

1	2	3	4	5	6	7	8	9	10
11	12	13	14	15	16	17	18	19	20
21	22	23	24	25	26	27	28	29	30
31	32	33	34	35	36	37	38	39	40
41	42	43		45	46	47	48	49	50

44

35

33

Directions: Circle the missing number. Count to 50 starting with that number. Color the boxes as you count.

Apply and Grow: Practice

1	2	3	4	5	6	7	8	9	10
11	12	13	14	15	16	17	18	19	20
	22	23	24	25	26	27	28	29	30
31	32	33	34	35	36	37	38	39	40
41	42	43	44	45	46	47	48	49	50

12

21

30

1	2	3	4	5	6	7	8	9	10
11	12	13	14	15	16	17	18	19	20
21	22	23	24	25	26	27	28	29	30
31	32	33	34	35	36				40
41	42	43	44	45	46	47	48	49	50

37, 38, 39 | 46, 47, 48 | 26, 27, 28

1	2	3	4	5	6	7	8	9	10
11	12	13	14	15	16	17	18	19	20
21	22	23	24	25	26	27	28	29	30
31	32	33	34	35	36	37	38	39	40
41	42	43	44	45	46	47	48	49	50

Directions: ❶ Circle the missing number. Count to 50 starting with that number. Color the boxes as you count. ❷ Circle the missing numbers. Tell how the missing numbers are alike. ❸ Find and circle the numbers *thirty-two* and *forty-seven*.

 **Think and Grow: Modeling Real Life**

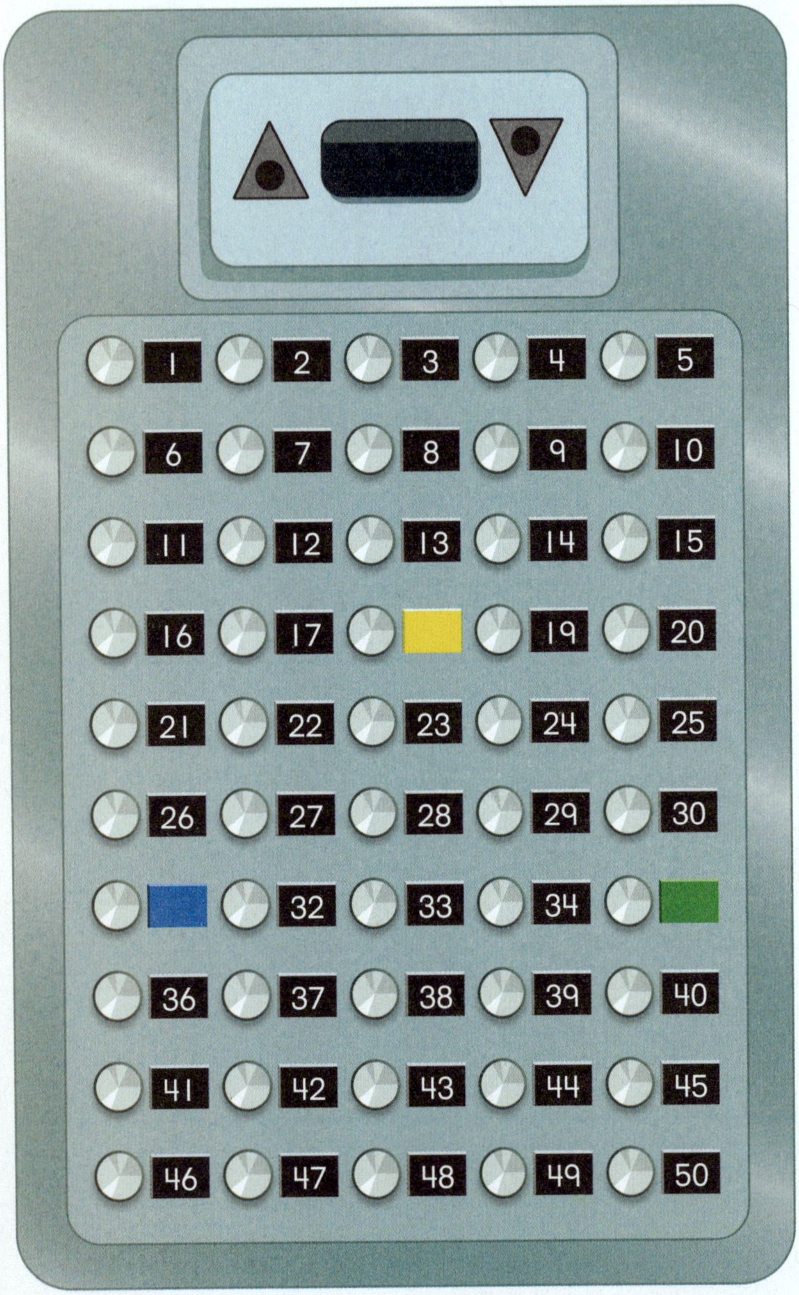

35

🟦 27

31

18

🟨 14

22

40

🟩 31

35

Directions: Circle the missing floor number for each color on the elevator keypad. Circle the color that covers the lowest missing floor number. Underline the color that covers the highest missing floor number.

Learning Target: Count to 50 by ones.

1	2	3	4	5	6	7	8	9	10
11	12	13	14	15	16	17	18	19	20
21	22	23	24	25	26	27	28	29	30
31	32	33	34	35		37	38	39	40
41	42	43	44	45	46	47	48	49	50

27

45

(36)

Directions: Circle the missing number. Count to 50 starting with that number. Color the boxes as you count.

 1

1	2	3	4	5	6	7	8	9	10
11	12	13	14	15	16	17	18	19	20
21	22	23	24	25	26	27	28	29	30
31	32	33		35	36	37	38	39	40
41	42	43	44	45	46	47	48	49	50

34

25

43

2

1	2	3	4	5	6	7	8	9	10
11	12	13	14	15	16	17	18	19	20
21	22	23	24	25	26	27	28	29	
31	32	33	34	35	36	37	38	39	40
41	42	43	44	45	46	47	48	49	50

21

30

39

Directions: **1** and **2** Circle the missing number. Count to 50 starting with that number. Color the boxes as you count.

1	2	3	4	5	6	7	8	9	10
11	12	13	14	15	16	17	18	19	20
21	22	23	24	25	26	27	28	29	30
31	32	33	34	35	36	37	38	39	40
			44	45	46	47	48	49	50

43, 44, 45 | 32, 33, 34 | 41, 42, 43

4

1	2	3	4	5	6	7	8	9	10
11	12	13	14	15	16	17	18	19	20
21	22	23	24	25	26	27	28	29	30
31	32	33	34	35	36	37	38	39	40
41	42	43	44	45	46	47	48	49	50

5

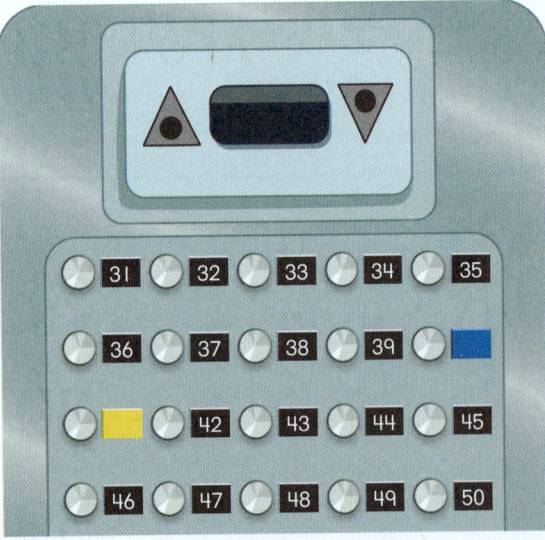

36   45

40   37

44   41

Directions: 3 Circle the missing numbers. Tell how the missing numbers are alike.
4 Find and circle the numbers *twenty-six* and *forty-two*. 5 Circle the missing floor
number for each color on the elevator keypad. Circle the color that covers the lowest
missing floor number. Underline the color that covers the highest missing floor number.

Name _____

 **Explore and Grow**

1	2	3	4	5	6	7	8	9	10
11	12	13	14	15	16	17	18	19	20
21	22	23	24	25	26	27	28	29	30
31	32	33	34	35	36	37	38	39	40
41	42	43	44	45	46	47	48	49	50
51	52	53	54	55	56	57	58	59	60
61	62	63	64	65	66	67	68	69	70
71	72	73	74	75	76	77	78	79	80
81	82	83	84	85	86	87	88	89	90
91	92	93	94	95	96	97	98	99	100

Directions: Point to each number as you count to 100. Color the numbers 30, 50, and 100.

1	2	3	4	5	6	7	8	9	10
11	12	13	14	15	16	17	18	19	20
21	22	23	24	25	26	27	28	29	30
31	32	33	34	35	⬤	37	38	39	40
41	42	43	44	45	46	47	48	49	50
51	52	53	54	55	56	57	58	⬤	60
61	62	63	64	65	66	67	68	69	70
71	72	⬤	74	75	76	77	78	79	80
81	82	83	84	85	86	87	88	89	90
91	92	93	94	95	96	97	98	99	100

27
(36)
45

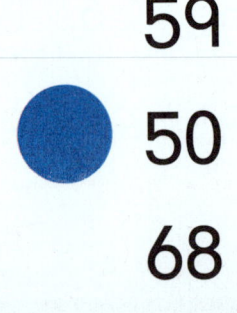

59
50
68

64
73
82

Directions: Circle the first missing number. Count to 100 starting with that number. Color the boxes as you count. Circle the other missing numbers as you count and color to 100.

512   five hundred twelve

✓ Apply and Grow: Practice

1	2	3	4	5	6	7	8	9	10
11	12	13	14	15	16	17	18	19	20
21	22	23	24	25	26	27	28	29	30
31	32	33	34	35	36	37	38	39	40
41	42	43	44	45	46	🔵	48	49	50
51	52	53	54	55	56	57	58	59	60
🔴	62	63	64	65	66	67	68	69	70
71	72	73	74	75	76	77	78	79	80
81	82	83	84	85	86	87	88	89	90
91	92	🟡	94	95	96	97	98	99	100

🍏 47
🔵 56
38

70
🔴 52
61

82
🟡 93
84

Directions: ① Circle the first missing number. Count to 100 starting with that number. Color the boxes as you count. Circle the other missing numbers as you count and color to 100.

Chapter 10 | Lesson 3

Think and Grow: Modeling Real Life

| 61 | 62 | 63 | 64 | 65 | 66 | | 68 | 69 | 70 |

| 71 | 72 | 73 | 74 | 75 | 76 | 77 | | 79 | 80 |

| 81 | 82 | 83 | 84 | 85 | | 87 | 88 | 89 | 90 |

| 78 |  | 67 |  | 86 |  |

Directions: Your friend, Newton, and Descartes each have 10 prize tickets.
They each lose 1 ticket.
- Circle the owner of each lost ticket.
- The winning ticket number is 1 more than 70. Circle the winning ticket number.
  Who is the winner? Circle the face of the winning ticket holder.

Learning Target: Count to 100 by ones.

1	2	3	4	5	6	7	8	9	10
11	12	13	14	15	16	17	18	19	20
21	22	23	24	25	26	27	28	29	30
31	32	33	34	35	36	37	38	39	40
41	42	43	44	45	46	47	48	49	50
51	52	53	54	55	●	57	58	59	60
61	62	63	64	65	66	67	68	69	70
71	72	73	74	75	76	77	●	79	80
81	82	83	84	85	86	87	88	89	90
91	92	93	94	95	96	97	98	99	100

●   65   (56)   47

●   55   68   (78)

Directions: Circle the missing number. Count to 100 starting with that number. Color the boxes as you count. Circle the other missing number as you count and color to 100.

1	2	3	4	5	6	7	8	9	10
11	12	13	14	15	16	17	18	19	20
21	22	23	24	25	26	27	28	29	30
31	32	33	34	35	36	37	38	39	40
41	42	43	44	45	46	47	48	49	50
51	52	●	54	55	56	57	58	59	60
61	62	63	64	65	66	67	68	69	70
71	72	73	74	75	76	77	78	79	80
81	82	●	84	85	86	87	88	89	90
91	92	93	94	95	96	97	98	99	100

🍏 ●   53   44   83   ┆   ●   74   92   83

Directions: 🍏 Circle the first missing number. Count to 100 starting with that number. Color the boxes as you count. Circle the other missing number as you count and color to 100.

Chapter 10 | Lesson 3

2

1	2	3	4	5	6	7	8	9	10
11	12	13	14	15	16	17	18	19	20
21	22	23	24	25	26	27	28	29	30
31	32	33	34	35	36	37	38	39	40
41	42	43	44	45	46	47	48	49	50
51	52	53	54	55	56	57	58	59	60
61	62	63	64	65	66	67	68	69	70
71	72	73	74	75	76	77	78		
	82	83	84	85	86	87	88	89	90
91	92	93	94	95	96	97	98	99	100

68, 69, 70 | 79, 80, 81 | 86, 87, 88

3

| 51 | 52 | | 54 | 55 | 56 | 57 | 58 | 59 | 60 |

| 61 | 62 | 63 | 64 | 65 | 66 | 67 | | 69 | 70 |

 53  | 68

Directions: Circle the missing numbers.  Newton and Descartes both have 10 prize tickets. They both lose 1 ticket. Circle the owner of each lost ticket. The winning ticket number is 1 more than 50. Circle the winning ticket number. Who is the winner? Circle the face of the winning ticket holder.

516   five hundred sixteen

Learning Target: Count to 100 by tens.

 **Explore and Grow**

1	2	3	4	5	6	7	8	9	10
11	12	13	14	15	16	17	18	19	20
21	22	23	24	25	26	27	28	29	30
31	32	33	34	35	36	37	38	39	40
41	42	43	44	45	46	47	48	49	50
51	52	53	54	55	56	57	58	59	60
61	62	63	64	65	66	67	68	69	70
71	72	73	74	75	76	77	78	79	80
81	82	83	84	85	86	87	88	89	90
91	92	93	94	95	96	97	98	99	100

Directions: Count to 10. Circle the number. Count 10 more. Circle the number. Repeat this process until you reach 100.

1	2	3	4	5	6	7	8	9	10
11	12	13	14	15	16	17	18	19	20
21	22	23	24	25	26	27	28	29	30
31	32	33	34	35	36	37	38	39	40
41	42	43	44	45	46	47	48	49	50
51	52	53	54	55	56	57	58	59	
61	62	63	64	65	66	67	68	69	70
71	72	73	74	75	76	77	78	79	80
81	82	83	84	85	86	87	88	89	90
91	92	93	94	95	96	97	98	99	100

20   40   (60)

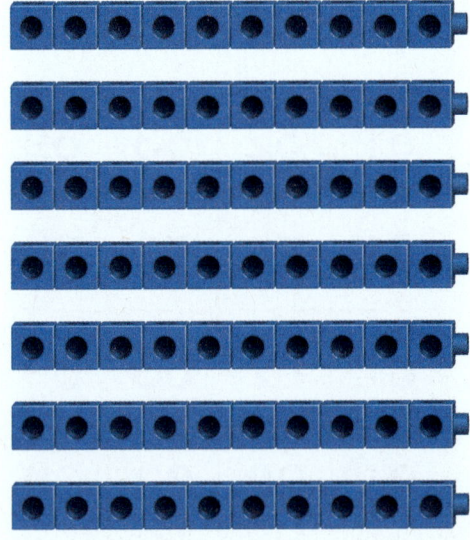

50   (60)   70

1	2	3	4	5	6	7	8	9	10
11	12	13	14	15	16	17	18	19	20
21	22	23	24	25	26	27	28	29	30
31	32	33	34	35	36	37	38	39	40
41	42	43	44	45	46	47	48	49	50
51	52	53	54	55	56	57	58	59	60
61	62	63	64	65	66	67	68	69	70
71	72	73	74	75	76	77	78	79	80
81	82	83	84	85	86	87	88	89	
91	92	93	94	95	96	97	98	99	100

60   70   90

70   80   90

Directions:
- Count to 100 by tens. Color the boxes as you count. Circle the missing decade number.
- Count the linking cubes. Circle the number that tells how many.

518   five hundred eighteen

✔ Apply and Grow: Practice

20   30   50

40   80   100

1	2	3	4	5	6	7	8	9	10
11	12	13	14	15	16	17	18	19	
21	22	23	24	25	26	27	28	29	
31	32	33	34	35	36	37	38	39	
41	42	43	44	45	46	47	48	49	50
51	52	53	54	55	56	57	58	59	60
61	62	63	64	65	66	67	68	69	70
71	72	73	74	75	76	77	78	79	80
81	82	83	84	85	86	87	88	89	90
91	92	93	94	95	96	97	98	99	100

15, 25, 35

20, 30, 40

40, 50, 60

Directions: 1 and 2 Count the objects. Circle the number that tells how many. 3 Count to 100 by tens. Color the boxes as you count. Circle the missing decade numbers.

10   20   30   40   50   60

10   20   30   40   50   60

Directions: Newton and Descartes each toss 6 balls. Each time a ball lands in their bucket, they earn 10 points. The player with the most points wins.

- Newton gets 3 balls in his bucket. Circle the number of points Newton earns. Draw a picture to show how you found your answer.

- Descartes gets 5 balls in his bucket. Circle the number of points Descartes earns. Draw a picture to show how you found your answer.

- Who earns more points? Circle the face of the winning player.

Learning Target: Count to 100 by tens.

1	2	3	4	5	6	7	8	9	10
11	12	13	14	15	16	17	18	19	20
21	22	23	24	25	26	27	28	29	30
31	32	33	34	35	36	37	38	39	
41	42	43	44	45	46	47	48	49	50
51	52	53	54	55	56	57	58	59	60
61	62	63	64	65	66	67	68	69	70
71	72	73	74	75	76	77	78	79	80
81	82	83	84	85	86	87	88	89	90
91	92	93	94	95	96	97	98	99	100

(40) 20 70

50 (40) 70

Directions:
- Count to 100 by tens. Color the boxes as you count. Circle the missing decade number.
- Count the linking cubes. Circle the number that tells how many.

30 60 50

50 65 70

Directions: and Count the objects. Circle the number that tells how many.

3

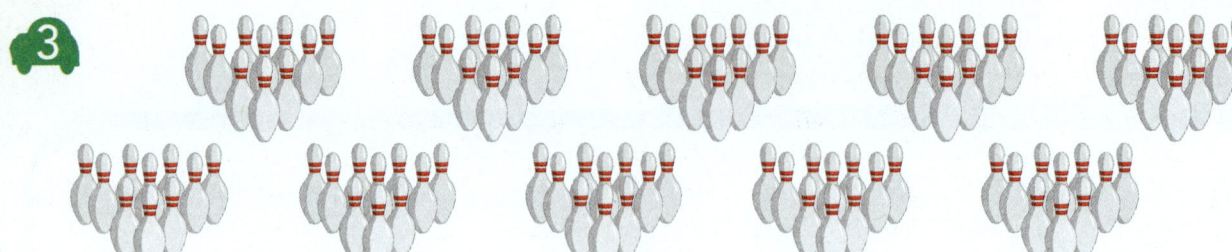

10   90   100

 4

1	2	3	4	5	6	7	8	9	10
11	12	13	14	15	16	17	18	19	20
21	22	23	24	25	26	27	28	29	30
31	32	33	34	35	36	37	38	39	40
41	42	43	44	45	46	47	48	49	50
51	52	53	54	55	56	57	58	59	60
61	62	63	64	65	66	67	68	69	
71	72	73	74	75	76	77	78	79	
81	82	83	84	85	86	87	88	89	
91	92	93	94	95	96	97	98	99	100

70, 75, 80

50, 40, 30

70, 80, 90

5

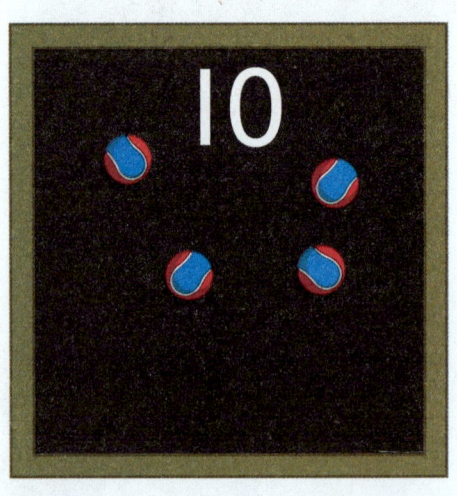

50

4

40

Directions: **3** Count the bowling pins. Circle the number that tells how many. **4** Count to 100 by tens. Color the boxes as you count. Circle the missing decade numbers. **5** Each time a ball sticks to the wall you earn 10 points. Four of your balls stick to the wall. Circle the number of points you earn. Draw a picture to show how you found your answer.

Name _____

Learning Target: Count by tens and ones within 100.

 **Explore and Grow**

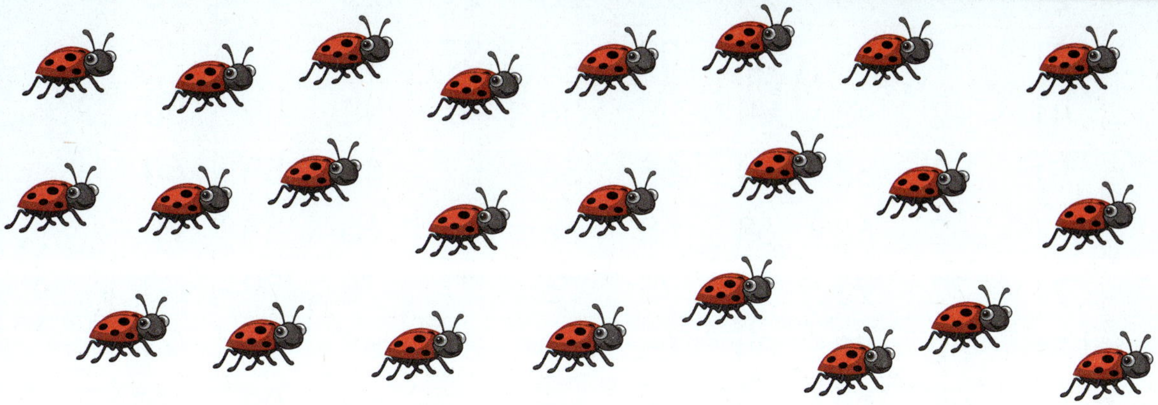

1	2	3	4	5	6	7	8	9	10
11	12	13	14	15	16	17	18	19	20
21	22	23	24	25	26	27	28	29	30
31	32	33	34	35	36	37	38	39	40
41	42	43	44	45	46	47	48	49	50
51	52	53	54	55	56	57	58	59	60
61	62	63	64	65	66	67	68	69	70
71	72	73	74	75	76	77	78	79	80
81	82	83	84	85	86	87	88	89	90
91	92	93	94	95	96	97	98	99	100

Directions: Circle groups of 10 ladybugs. Count the ladybugs. Circle the number in the chart that tells how many. Color to show how you counted.

Think and Grow

1	2	3	4	5	6	7	8	9	10
11	12	13	14	15	16	17	18	19	20
21	22	23	24	25	26	27	28	29	30
31	32	33	34	35	36	37	38	39	40
41	42	43	44	45	46	47	48	49	50
51	52	53	54	55	56	57	58	59	60
61	62	63	64	65	66	67	68	69	70
71	72	73	74	75	76	77	78	79	80
81	82	83	84	85	86	87	88	89	90
91	92	93	94	95	96	97	98	99	100

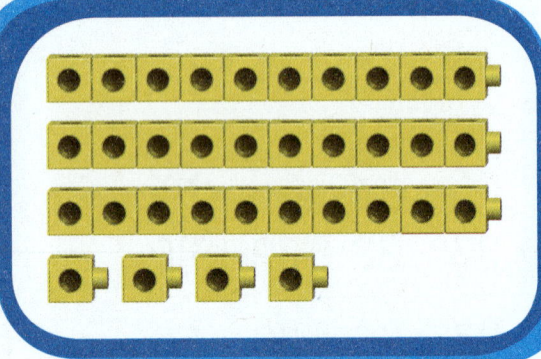

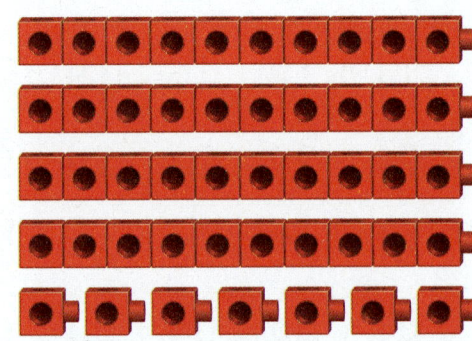

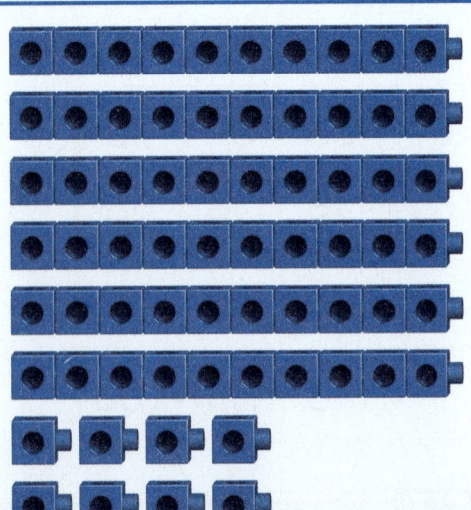

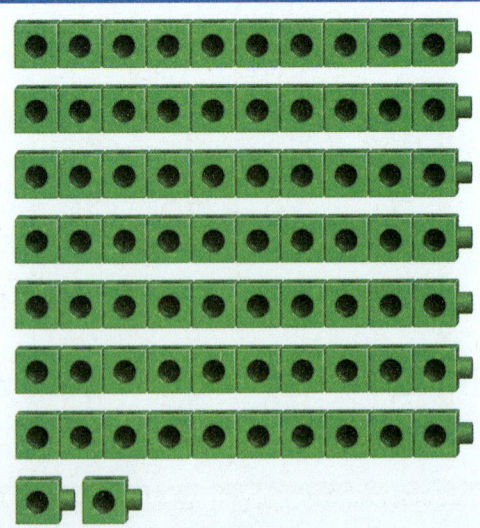

Directions: Count the linking cubes. Find the number in the hundred chart that tells how many. Color the number. Tell how you counted.

Name _____

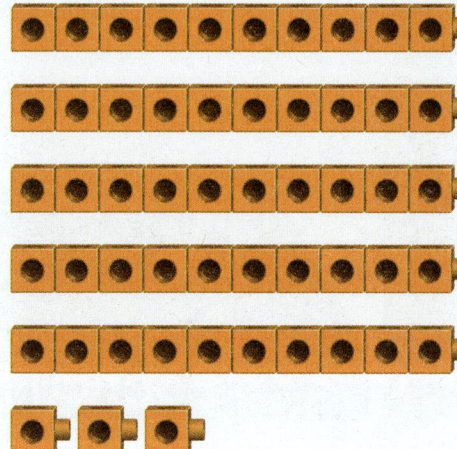

43

35

53

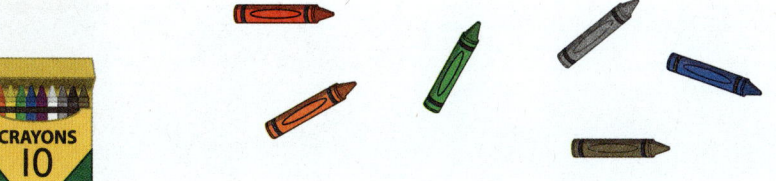

29

11

38

84

48

12

Directions: ① – ③ Count the objects. Circle the number that tells how many. Tell how you counted.

6   14   59   60   95

8   20   61   79   80

Directions: A toy store sells bouncy balls individually and in bags of 10.
- Count the bouncy balls. Circle the number that tells how many.
- The store orders 2 more bags of bouncy balls. Draw the new bags of balls.
- Circle the number that tells how many bouncy balls the store has now.

Learning Target: Count by tens and ones within 100.

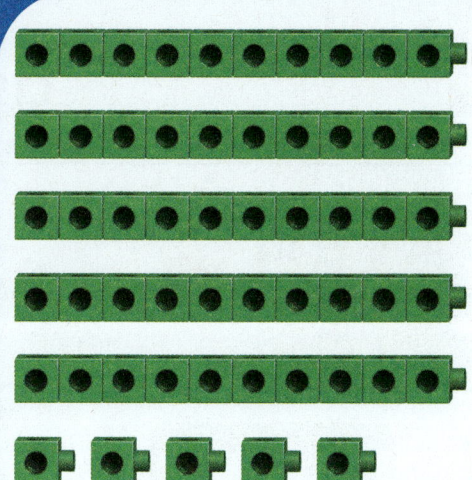

1	2	3	4	5	6	7	8	9	10
11	12	13	14	15	16	17	18	19	20
21	22	23	24	25	26	27	28	29	30
31	32	33	34	35	36	37	38	39	40
41	42	43	44	45	46	47	48	49	50
51	52	53	54	55	56	57	58	59	60
61	62	63	64	65	66	67	68	69	70
71	72	73	74	75	76	77	78	79	80
81	82	83	84	85	86	87	88	89	90
91	92	93	94	95	96	97	98	99	100

by tens and by ones

Directions: Count the linking cubes. Find the number in the chart that tells how many. Color the number. Tell how you counted.

1

10

19

20

2

11

47

56

Directions: **1** and **2** Count the objects. Circle the number that tells how many. Tell how you counted.

29

90

92

36

9

63

9   66

21   62

32   39

36   12

Directions: and Count the objects. Circle the number that tells how many. Tell how you counted. Count the bottles of bubbles. Circle the number that tells how many. Draw 3 more packs of bubbles. Circle the number that tells how many bottles of bubbles there are now.

528   five hundred twenty-eight

Name _____

Learning Target: Count by tens from a given number within 100.

 **Explore and Grow**

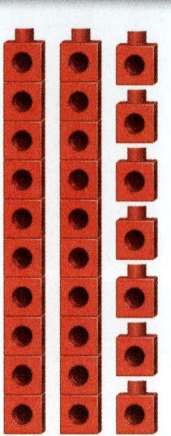

1	2	3	4	5	6	7	8	9	10
11	12	13	14	15	16	17	18	19	20
21	22	23	24	25	26	27	28	29	30
31	32	33	34	35	36	37	38	39	40
41	42	43	44	45	46	47	48	49	50
51	52	53	54	55	56	57	58	59	60
61	62	63	64	65	66	67	68	69	70
71	72	73	74	75	76	77	78	79	80
81	82	83	84	85	86	87	88	89	90
91	92	93	94	95	96	97	98	99	100

Directions: Count each group of linking cubes. Circle the numbers in the chart that tell how many for each group. What is the same in each number? What is different in each number?

1	2	3	(4)	5	6	7	8	9	10
11	12	13		15	16	17	18	19	20
21	22	23		25	26	27	28	29	30
31	32	33		35	36	37	38	39	40
41	42	43	44	45	46	47	(48)	49	50
(51)	52	53	54	55	56	57		59	60
	62	63	64	65	66	67		69	70
	72	73	74	75	76	77		79	80
	82	83	84	85	86	87	88	89	90
91	92	93	94	95	96	97	98	99	100

(51)   52, 53, 54 | 61, 62, 63 | ⟨61, 71, 81⟩

(4)   14, 24, 34 | 5, 6, 7 | 14, 15, 16

(48)   38, 28, 18 | 58, 68, 78 | 49, 50, 51

Directions: Count by tens starting with the circled number. Circle the correct group of missing numbers.

 **Apply and Grow: Practice**

1	2	3	4	5	6	7	8	9	10
11	12	13	14	15	16	(17)	18	19	20
21	22	23	24	25	26		28	29	30
31	32	33	(34)	35	36		38	39	40
41	42	43		45	(46)		48	49	50

1 (17)   18, 19, 20 | 26, 36, 46 | 27, 37, 47

2 (34)   23, 33, 43 | 44, 54, 64 | 35, 36, 37

3 (46)   56, 66, 76 | 47, 48, 49 | 26, 36, 46

4   19, 29, 39, 49, 59, ____, 79, 89, 99

  60   69   78

Directions: **1**–**3** Count by tens starting with the circled number. Circle the numbers that tell how you counted. **4** Circle the missing number. Tell how you counted.

5   7   20   25   50

Directions: A grocery store sells milk cartons individually and in boxes of 10.

• Count the milk cartons. Circle the number that tells how many.

• You need 75 milk cartons in all. Draw to show how many more boxes of milk cartons you need.

Learning Target: Count by tens from a given number within 100.

1	2	3	4	5	6	7	8	9	10
11	12	13	14	(15)	16	17	18	19	20
21	22	23	24		26	27	28	29	30
31	32	33	34		36	37	38	39	40
41	42	43	44		46	47	48	49	50

(25, 35, 45) | 16, 17, 18 | 4, 14, 24

Directions: Count by tens starting with the circled number. Circle the missing numbers.

1	2	3	4	5	6	7	8	9	10
(11)	12	13	14	15	16	17	18	19	20
	22	23	24	25	26	27	28	29	30
	32	33	34	35	36	37	38	39	40
	42	43	44	45	46	47	48	49	50

① 12, 13, 14 | 21, 31, 41 | 20, 30, 40

Directions: ① Count by tens starting with the circled number. Circle the missing numbers.

2 (62) 63, 64, 65 | 72, 82, 92 | 61, 71, 81

3 (19) 29, 39, 49 | 20, 21, 22 | 20, 30, 40

4 (53) 54, 55, 56 | 55, 57, 59 | 63, 73, 83

5 27, 37, 47, 57, 67, 77, ____, 97

78   96   87

6

16

26

10

Directions: **2**–**4** Count by tens starting with the circled number. Circle the numbers that tell how you counted. **5** Circle the missing number. Tell how you counted. **6** Count the cups of applesauce. Circle the number that tells how many. You need 36 cups of applesauce in all. Draw to show how many more boxes you need.

 1

65

11

56

 2

 3

59			
77	66, 76, 86	57, 58, 59	57, 67, 77
86			

Directions: A store sells gift bags and party blowers in packages of 10 and individually. 1 Count the gift bags. Circle the number that tells how many. 2 You buy an equal number of party blowers and gift bags. Draw more party blowers to complete the picture. 3 You buy 3 more packages of party blowers for the adults. Circle the number that tells how many party blowers there are in all. Circle the group of numbers that tells how you counted.

Chapter 10

Hundred Chart Puzzle

Directions: Cut out the Hundred Chart Puzzle Pieces. Put the pieces together to complete the hundred chart.

10.1 **Count to 30 by Ones**

1	2	3	4	5	6	7	8	9	10
11	12	13	14	15	16		18	19	20
21	22	23	24	25	26	27	28	29	30

8

17

26

1	2	3	4	5	6	7	8	9	10
11	12	13	14	15	16	17	18	19	20
21	22	23	24	25	26	27	28	29	30

10.2 **Count to 50 by Ones**

3

1	2	3	4	5	6	7	8	9	10
11	12	13	14	15	16	17	18	19	20
			24	25	26	27	28	29	30
31	32	33	34	35	36	37	38	39	40
41	42	43	44	45	46	47	48	49	50

21, 22, 23 | 28, 29, 30 | 12, 13, 14

Directions: ❶ Circle the missing number. Count to 30 starting with that number. Color the boxes as you count. ❷ Find and circle the number *twenty-nine*. ❸ Circle the missing numbers. Tell how the missing numbers are alike.

Count to 100 by Ones

1	2	3	4	5	6	7	8	9	10
11	12	13	14	15	16	17	18	19	20
21	●	23	24	25	26	27	28	29	30
31	32	33	34	35	36	37	38	39	40
41	42	43	44	45	46	47	48	49	50
51	52	53	54	●	56	57	58	59	60
61	62	63	64	65	66	67	68	69	70
●	72	73	74	75	76	77	78	79	●
81	82	83	84	85	86	87	88	89	90
91	92	93	94	95	96	97	98	99	100

	22		46		62		71
🟡	13	🔵(navy)	64	🔵	71	🔴	80
	31		55		80		89

Directions: 🐸 Circle the first missing number. Count to 100 starting with that number. Color the boxes as you count. Circle the other missing numbers as you count and color to 100.

Count to 100 by Tens

40   65   70

60   70   90

10.5 **Count by Tens and Ones**

18   88   99

Directions: ⭐ and 🍀 Count the objects. Circle the number that tells how many.
💚 Count the crayons. Circle the number that tells how many. Tell how you counted.

8

7		90
34		72
70		54

10.6 **Count by Tens from a Number**

9 ⑨ 19, 29, 39 | 10, 20, 30 | 62, 73, 84

---

10 ㊻ 29, 39, 49 | 56, 66, 76 | 3, 40, 51

11

	32
	42
	50

Directions: **8** Count the apples. Circle the number that tells how many. Draw 2 more bags of apples. Circle the number that tells how many apples there are now. **9** and **10** Count by tens starting with the circled number. Circle the numbers that tell how you counted. **11** Count the water bottles. Circle the number that tells how many. You need 42 water bottles in all. Draw to show how many more cases you need.

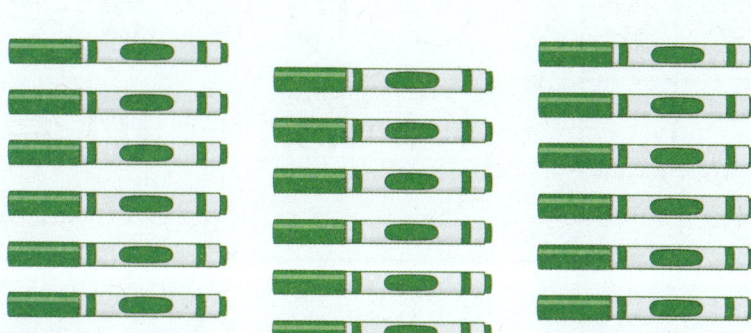

○ 17
○ 18
○ 19
○ 20

○ 60   ○ 90   ○ 80   ○ 100

$3 + 1 = ?$

○ 3
○ 1
○ 2
○ 4

Directions: Shade the circle next to the answer. ❶ and ❷ Which number tells how many? ❸ Which number completes the addition sentence?

4

○ $15 = 10 + 5$

○ $15 - 10 = 5$

○ $10 + 5 = 15$

○ $5 + 10 = 15$

5

7

○ 9

○ 8

○ 6

○ 10

6

○ 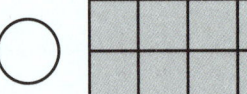   ○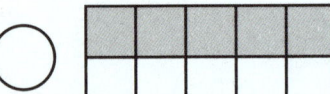

○ 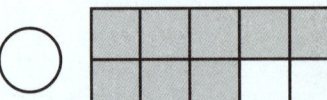   ○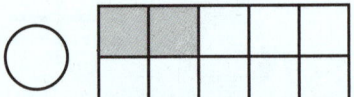

Directions: Shade the circle next to the answer. **4** Which number sentence does *not* tell how many flowers there are in all? **5** Which number is *not* greater than 7? **6** Which ten frame shows the number of tomatoes?

542   five hundred forty-two

7

_____

- - - - - - - - - -

_____

8

61	62	63	64	65	66	67	68	69	70
71	72	73	74	75	76	77	78	79	80
81	82	83		85	86	87	88	89	90
91	92	93	94	95	96	97	98	99	100

93

75

84

_____

- - - - - - - - - -

_____

Directions: 7 Count the peaches. Say the number. Write the number. 8 Circle the missing number. Count to 100 starting with that number. Color boxes as you count. Then circle a number in the chart that is greater than the missing number. 9 Draw 19 leaves on the ground. Write the number.

9   14

- - -   ▬▬ / ▬▬

    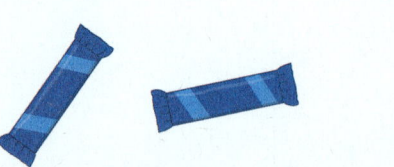

8

9

18

Directions: Count forward from 9 and stop at 14. Write the numbers you count. ⌂ Take apart the linking cubes. Circle the parts. Then write a subtraction sentence by taking one of the parts from the whole. 🍃 Count the granola bars. Circle the number that tells how many. You need 48 granola bars in all. Draw to show how many more boxes you need.

544   five hundred forty-four

Identify Two-Dimensional Shapes

- Do you have any pets?
- What shapes can you see in the picture?

© Big Ideas Learning, LLC

Name _____

Vocabulary

Review Words
classify
category
less than

spots   no spots

_____   _____

- - - - - -   - - - - - -

_____   _____

Directions: Classify the animals into the categories shown. Write the marks in the chart. Count the marks and write the numbers to tell how many animals are in each category. Draw a line through the number that is less than the other number.

Chapter 11 Vocabulary Cards

circle

curve

hexagon

rectangle

side

sort

square

triangle

© Big Ideas Learning, LLC

© Big Ideas Learning, LLC

© Big Ideas Learning, LLC

© Big Ideas Learning, LLC

© Big Ideas Learning, LLC

© Big Ideas Learning, LLC

© Big Ideas Learning, LLC

© Big Ideas Learning, LLC

Chapter 11 Vocabulary Cards

two-dimensional shape

vertex

vertices

Name _____

Learning Target: Describe
two-dimensional shapes.

 **Explore and Grow**

curves	no curves

Directions: Cut out the Two-Dimensional Shape Cards. Sort the cards into the
categories shown.

© Big Ideas Learning, LLC

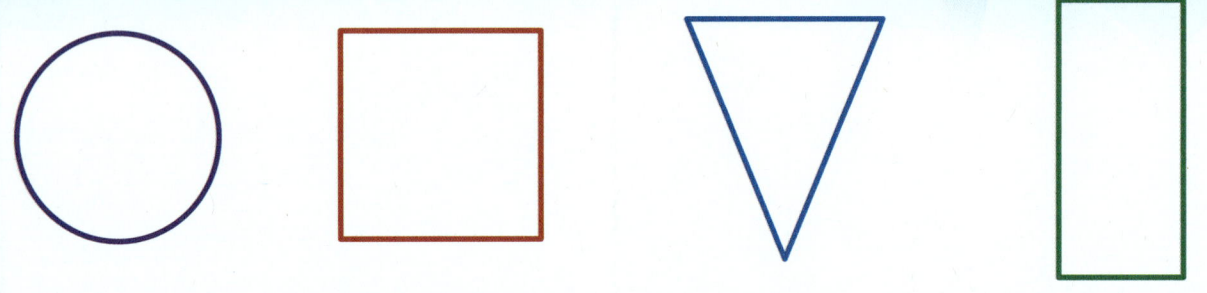

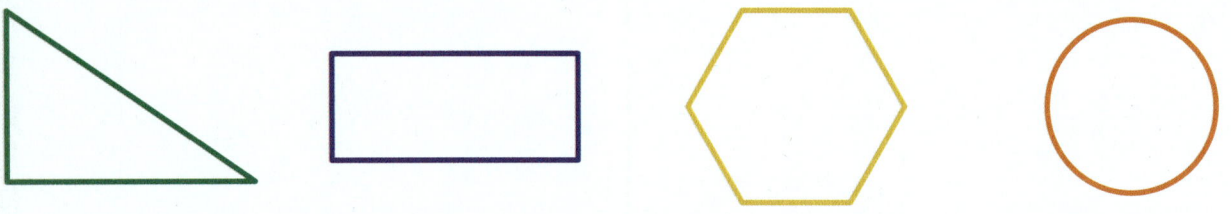

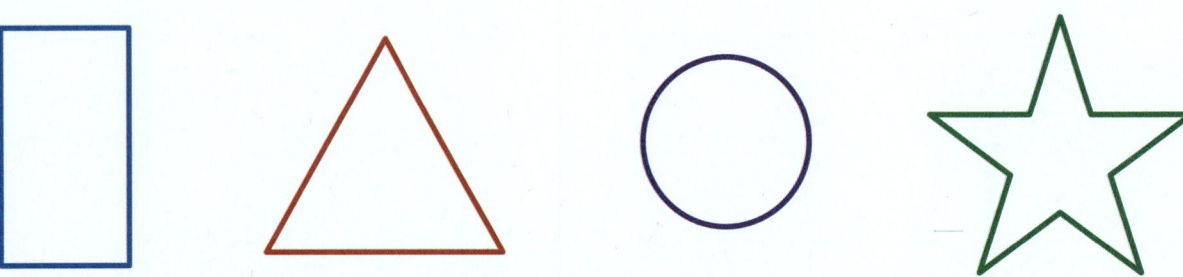

Directions:

- Color the shape that has a curve.
- Color the shape that does not have any vertices.
- Color the shape that has more than 4 vertices.
- Color the shape that has only 3 sides.

 **Apply and Grow: Practice**

 1

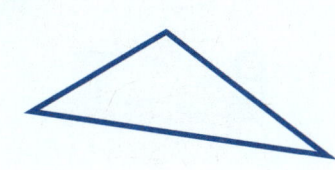 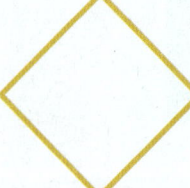 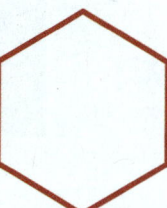 

 2

 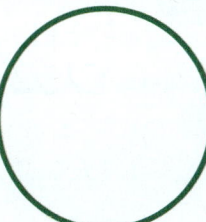  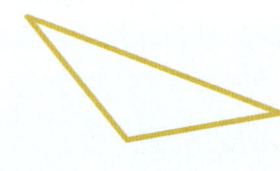

 3

 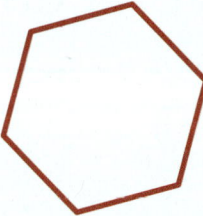 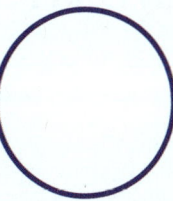 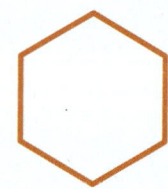

 4

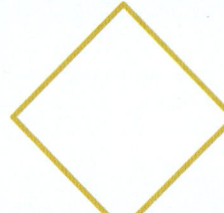  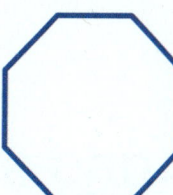 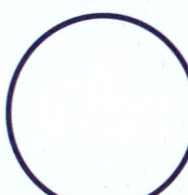

Directions: **1** Color the shapes that have only 4 sides. **2** Color the shapes that have only 3 vertices. **3** Color the shapes that have 6 vertices. **4** Color the shapes that have curves.

Chapter 11 | **Lesson 1**

five hundred forty-nine 549

more than 4 vertices

all straight sides

curves and straight sides

Directions: Write the number that answers the question.
• How many stickers in the picture have more than 4 vertices?
• How many stickers in the picture have all straight sides?
• How many stickers in the picture have both curves and straight sides?

Learning Target: Describe two-dimensional shapes.

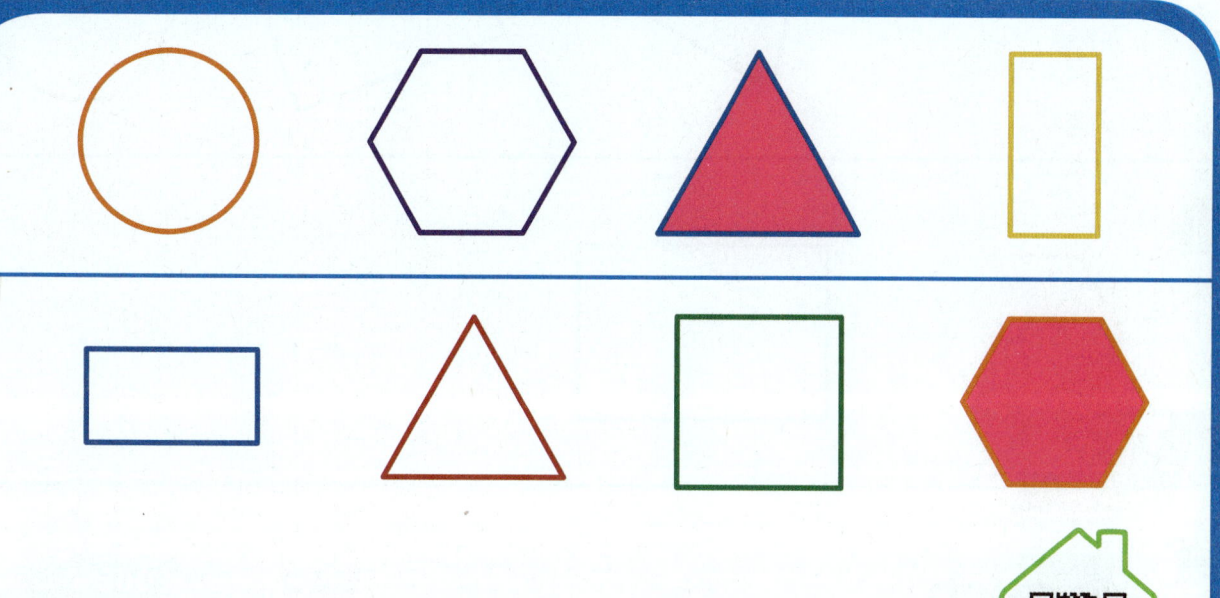

Directions:
- Color the shape that has only 3 vertices.
- Color the shape that has more than 4 sides.

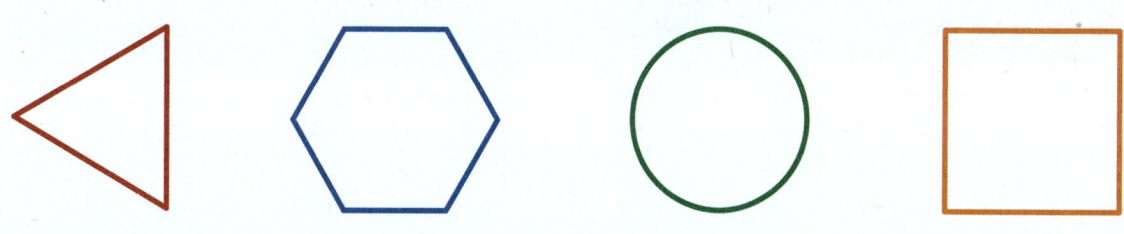

Directions: ① Color the shape that has 6 sides. ② Color the shape that has more than 3 straight sides.

3

4

5

less than 4 vertices _____

more than 3 sides _____

Directions: **3** Color the shapes that have curves. **4** Color the shapes that have only 4 vertices. **5** How many stickers have less than 4 vertices? Write the number. How many stickers have more than 3 sides? Write the number.

552   five hundred fifty-two

Learning Target: Identify and describe triangles.

 **Explore and Grow**

triangle	not a triangle

Directions: Cut out the Triangle or Not a Triangle Sort Cards. Sort the cards into the categories shown.

Think and Grow

3 sides

3 vertices

Directions: Color any triangles. Tell why your answers are correct.

Name _____

1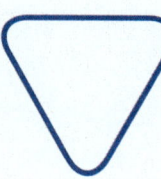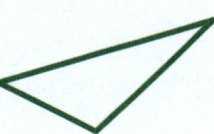

2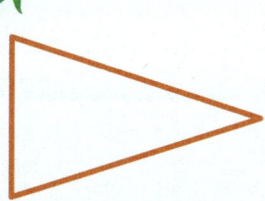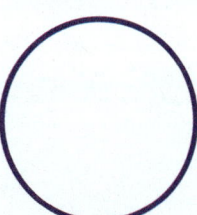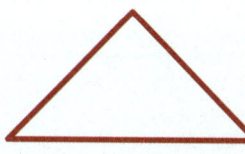

3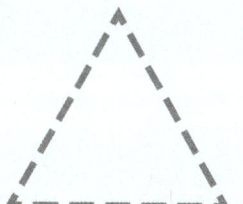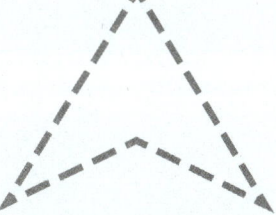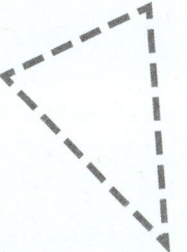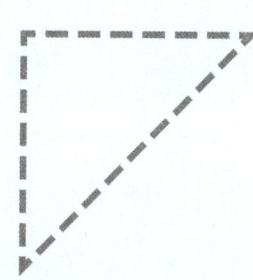

_____   _____

_ _ _ _ _ _ _ _ _ _ _ _   _ _ _ _ _ _ _ _ _ _ _ _

_____ sides   _____ vertices

Directions: **1** and **2** Color any triangles. Tell why your answers are correct.
3 Trace the shapes that are triangles. Write the number of sides and the number of vertices of a triangle.

Think and Grow: Modeling Real Life

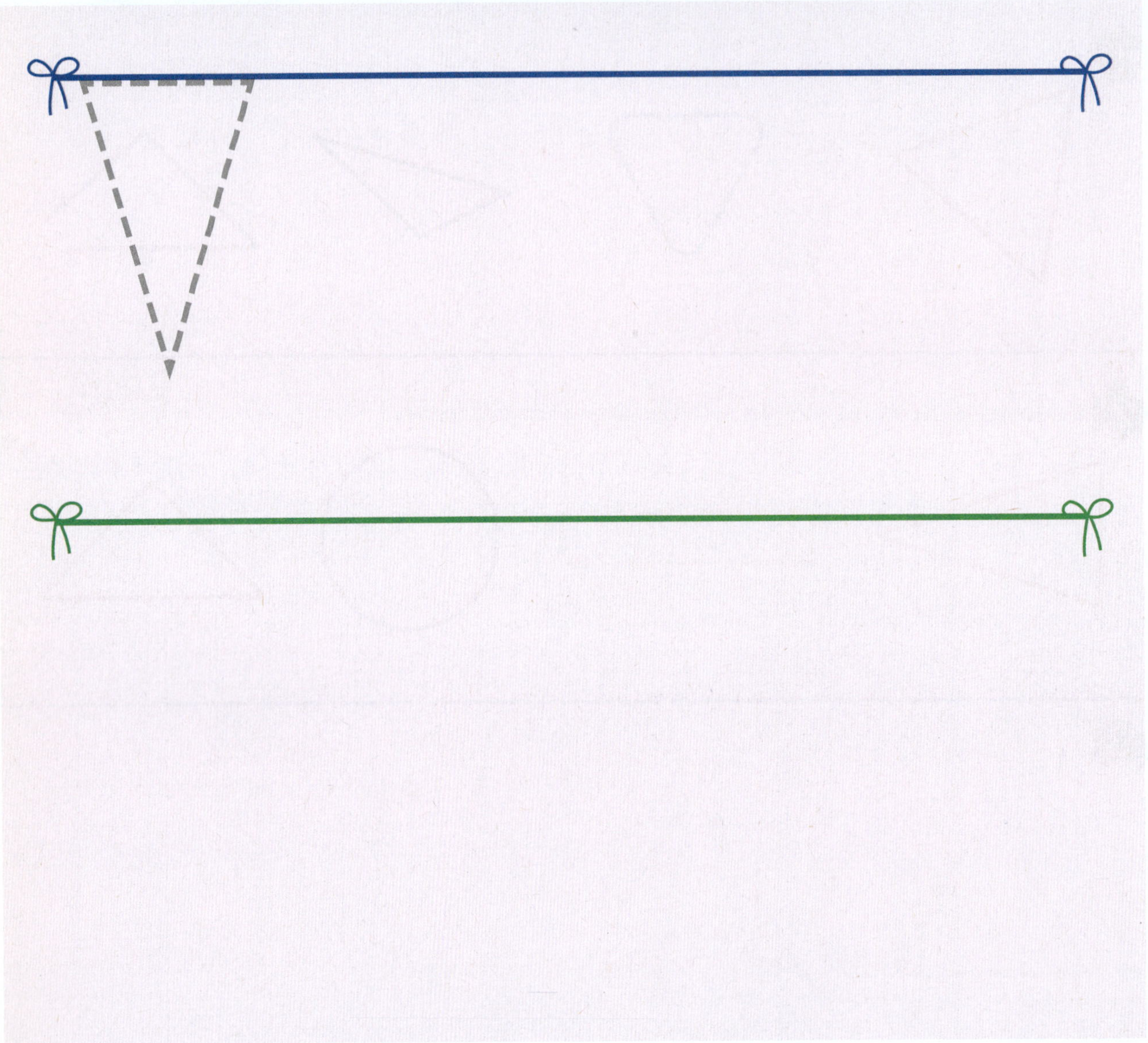

Directions: You use triangle-shaped flags to make two banners for a party. You use 10 flags in all.

- Draw and color flags to make the banners.
- Write an addition sentence to match your picture.

Name _____

Learning Target: Identify and describe triangles.

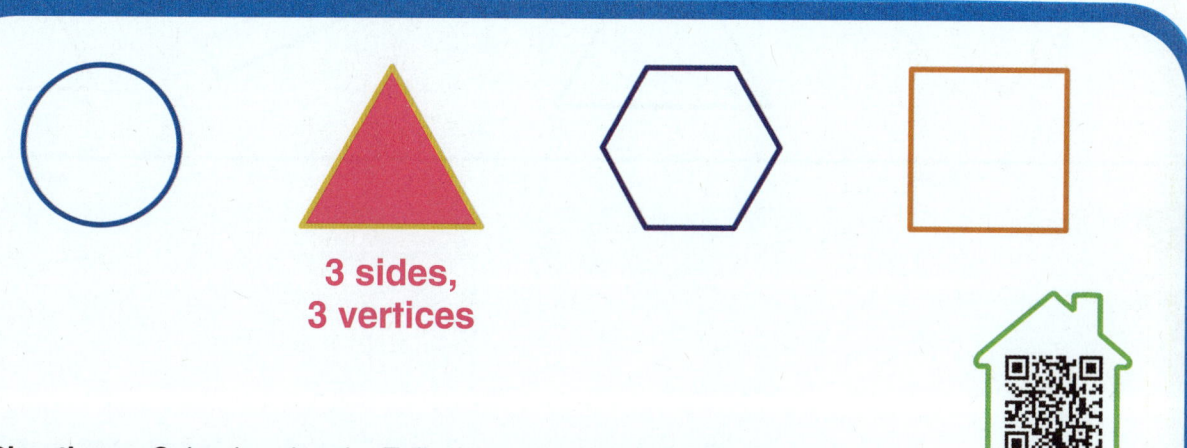

3 sides,
3 vertices

Directions: Color the triangle. Tell why your answer is correct.

1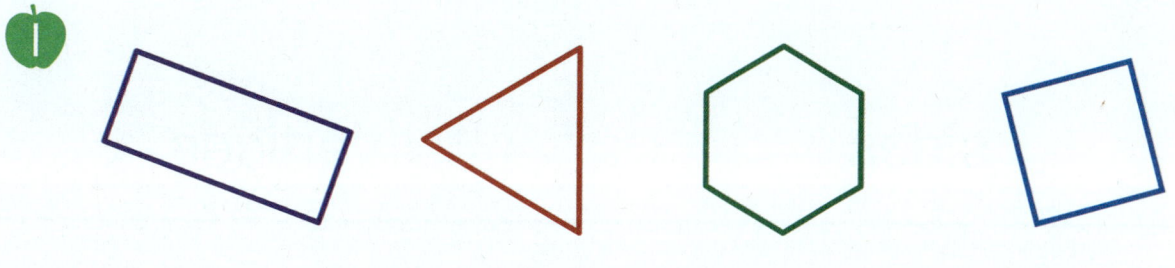

2

3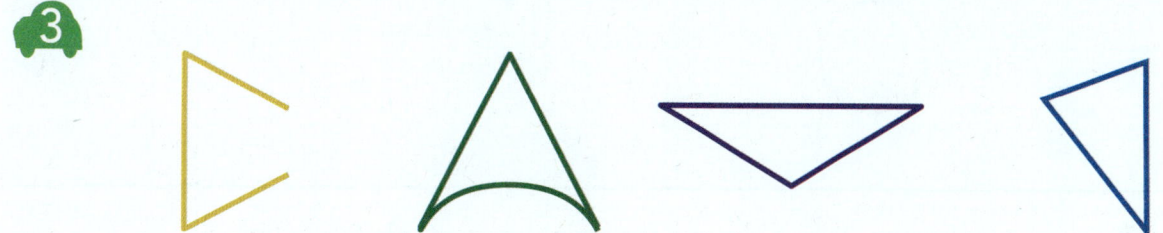

Directions: **1** – **3** Color any triangles. Tell why your answers are correct.

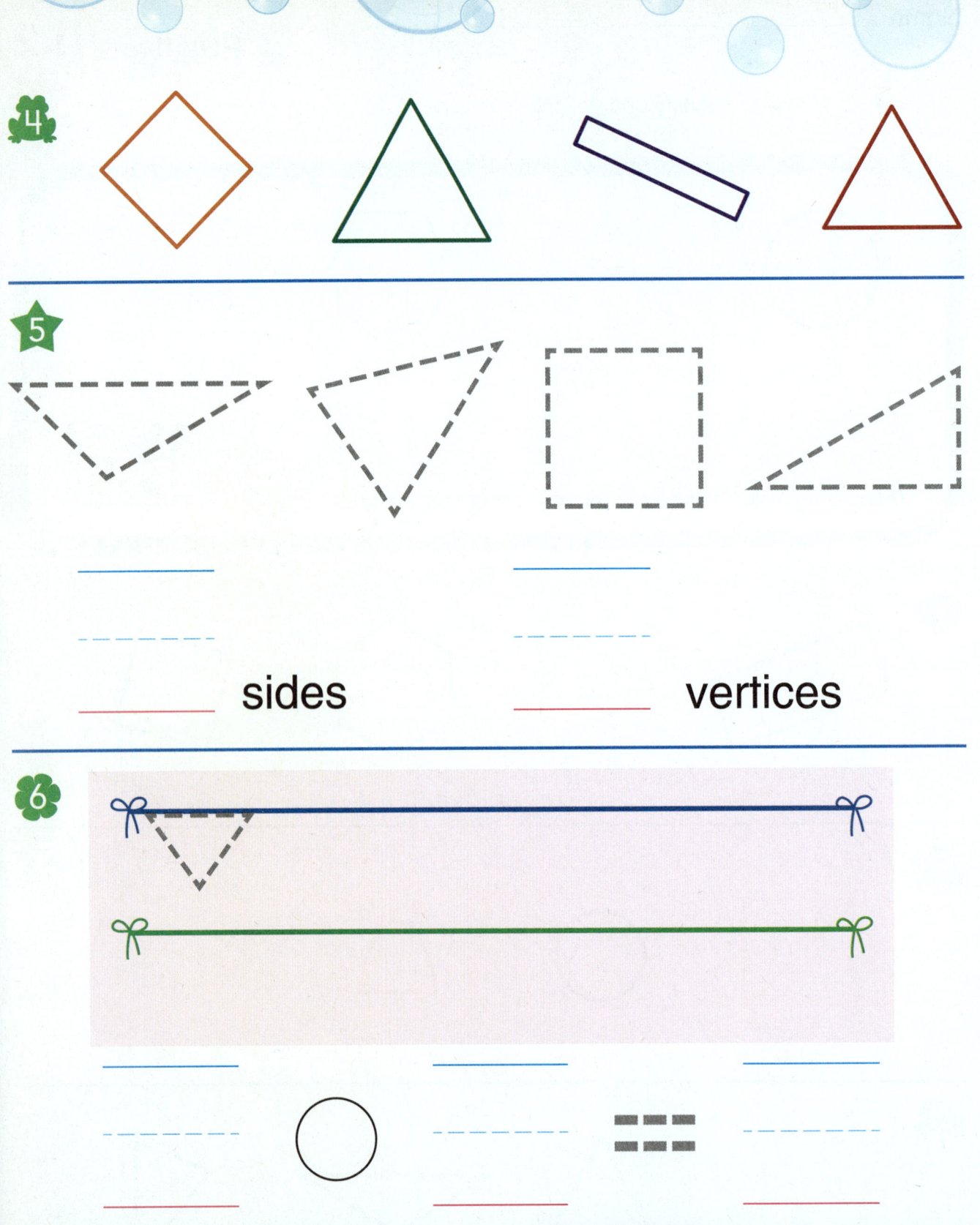

_____ sides   _____ vertices

Directions: 🐸 Color any triangles. Tell why your answers are correct. ⭐ Trace the shapes that are triangles. Write the number of sides and the number of vertices of a triangle. 🍀 You use triangle-shaped flags to make two banners for a party. You use 8 flags in all. Draw and color flags to make the banners. Then write an addition sentence to match your picture.

Learning Target: Identify and describe rectangles.

 **Explore and Grow**

rectangle	not a rectangle

Directions: Cut out the Rectangle or Not a Rectangle Sort Cards. Sort the cards into the categories shown.

Chapter 11 | **Lesson 3**

five hundred fifty-nine 559

© Big Ideas Learning, LLC

Think and Grow

4 sides, 4 vertices

L-shaped vertices

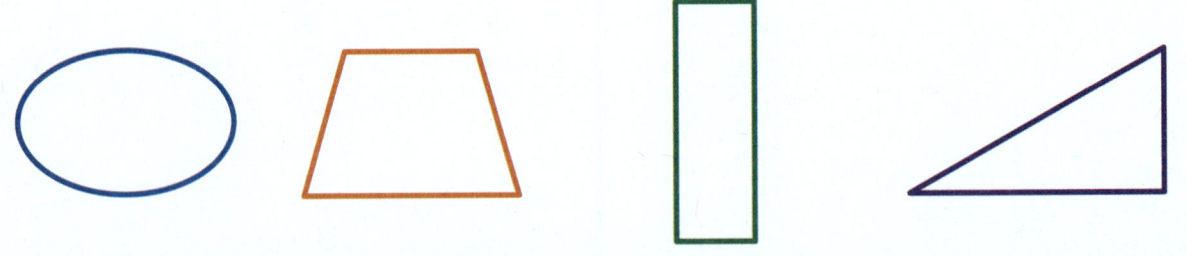

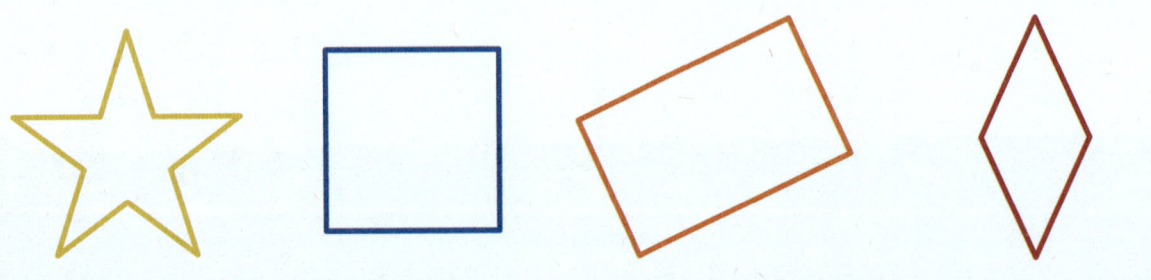

Directions: Color any rectangles. Tell why your answers are correct.

Name _____

1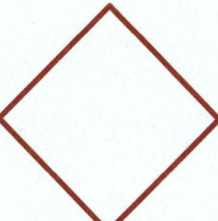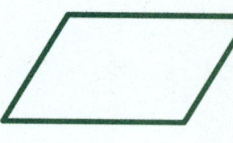

2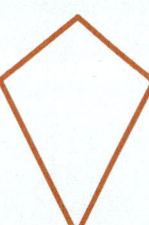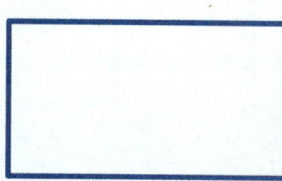

3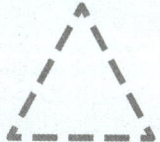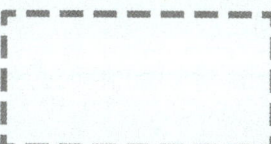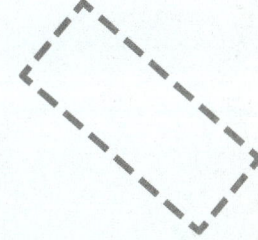

_____ sides   _____ vertices

Directions: **1** and **2** Color any rectangles. Tell why your answers are correct.
3 Trace the shapes that are rectangles. Write the number of sides and the
number of vertices of a rectangle.

Think and Grow: Modeling Real Life

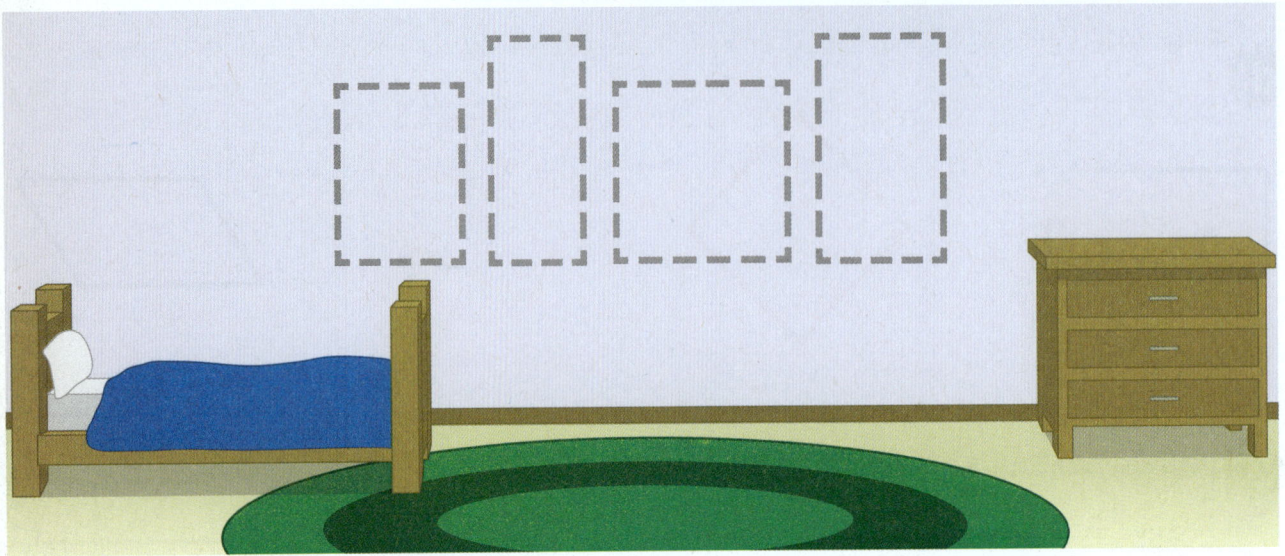

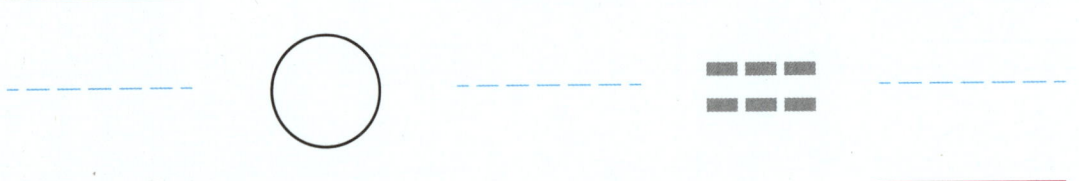

Directions:

- Trace and color 4 rectangular posters on the wall.
- You put 3 more posters on another wall. Draw and color 3 rectangular posters on the wall.
- Write an addition sentence to tell how many posters are on the walls in all.

562   five hundred sixty-two

Learning Target: Identify and describe rectangles.

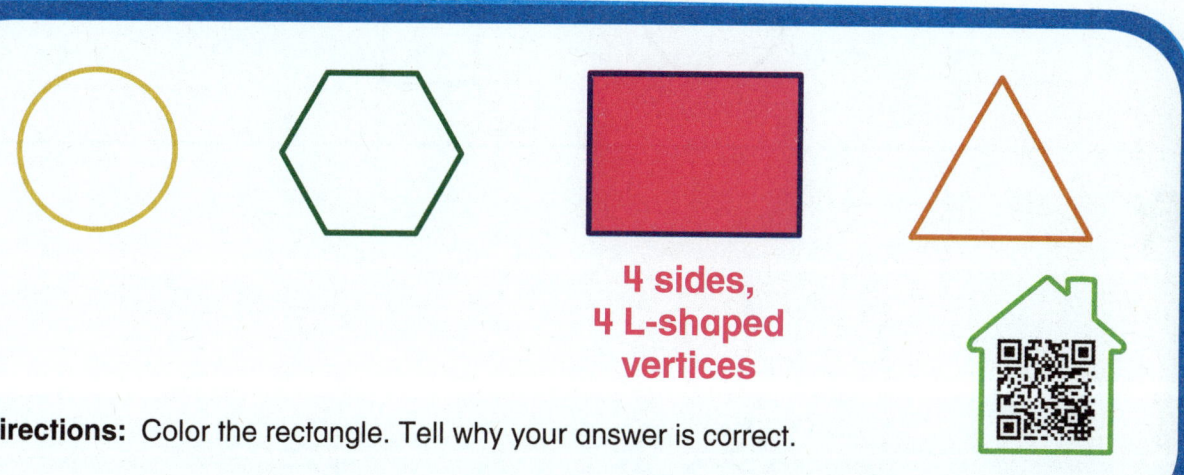

4 sides,
4 L-shaped
vertices

Directions: Color the rectangle. Tell why your answer is correct.

Directions: ➊–➌ Color any rectangles. Tell why your answers are correct.

Chapter 11 | Lesson 3

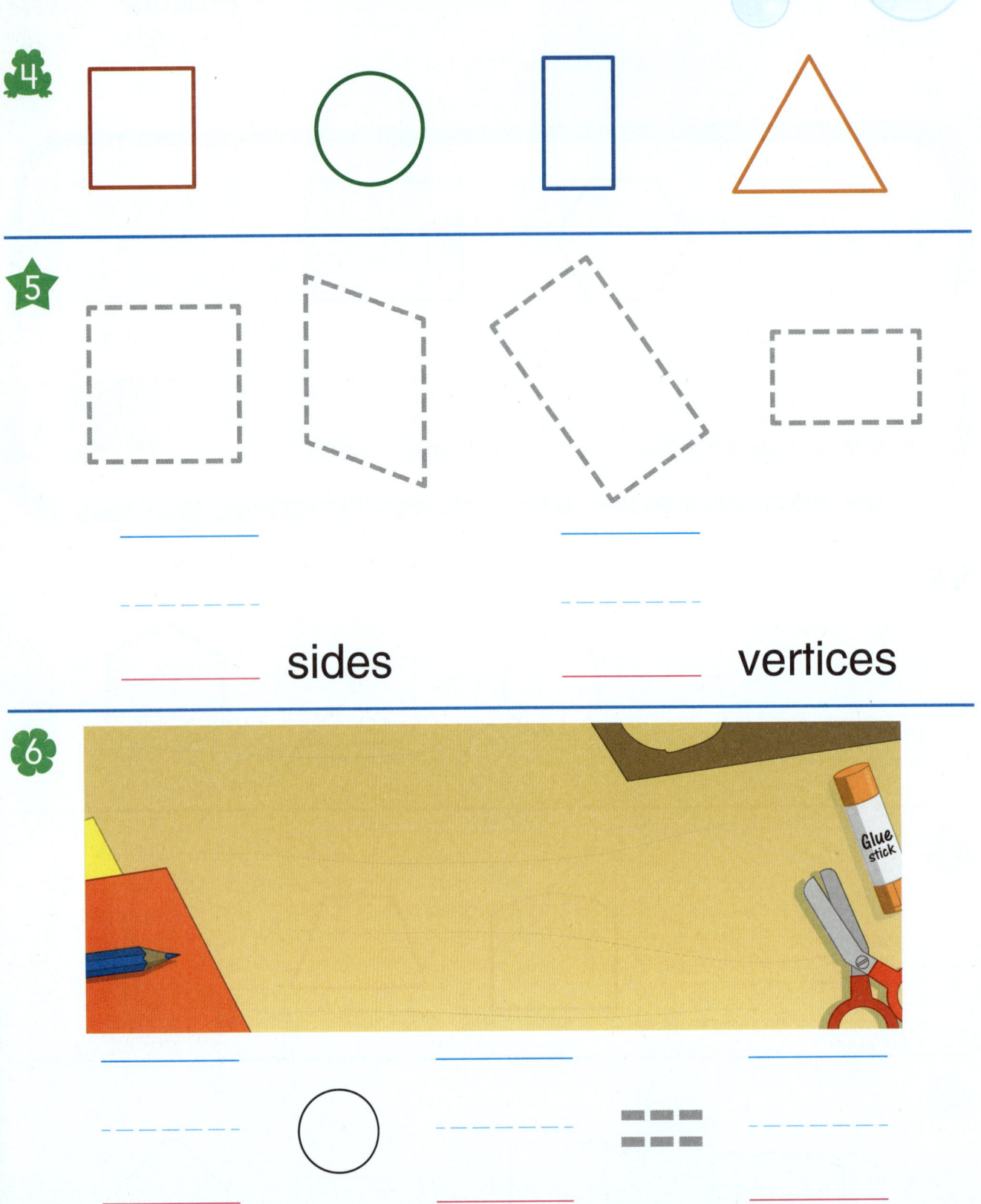

_____ sides   _____ vertices

Directions: 🐸 Color any rectangles. Tell why your answers are correct. ⭐ Trace the shapes that are rectangles. Write the number of sides and the number of vertices of a rectangle. 🍀 You make rectangular picture frames. Draw 1 large frame and 4 small frames. Then write an addition sentence to tell how many frames you make in all.

Name _____

Learning Target: Identify and describe squares.

 Explore and Grow

square	not a square

Directions: Cut out the Square or Not a Square Sort Cards. Sort the cards into the categories shown.

Think and Grow

4 sides
of same length,
4 L-shaped vertices

Special
rectangle

Directions: Color any squares. Tell why your answers are correct.

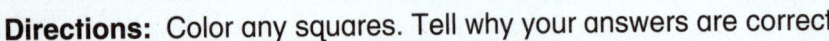

Name _____

 ①

 ②

  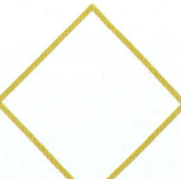 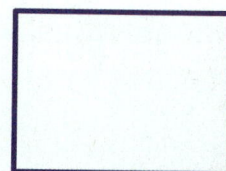

 ③

 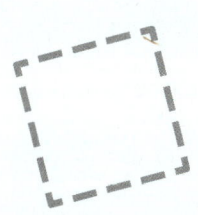 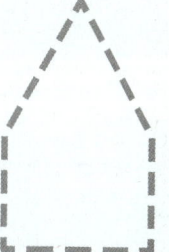 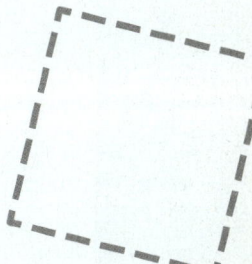

_____ _____

_ _ _ _ _ _ _   _ _ _ _ _ _ _

_____ sides   _____ vertices

Directions: ① and ② Color any squares. Tell why your answers are correct.
③ Trace the shapes that are squares. Write the number of sides and the number of vertices of a square.

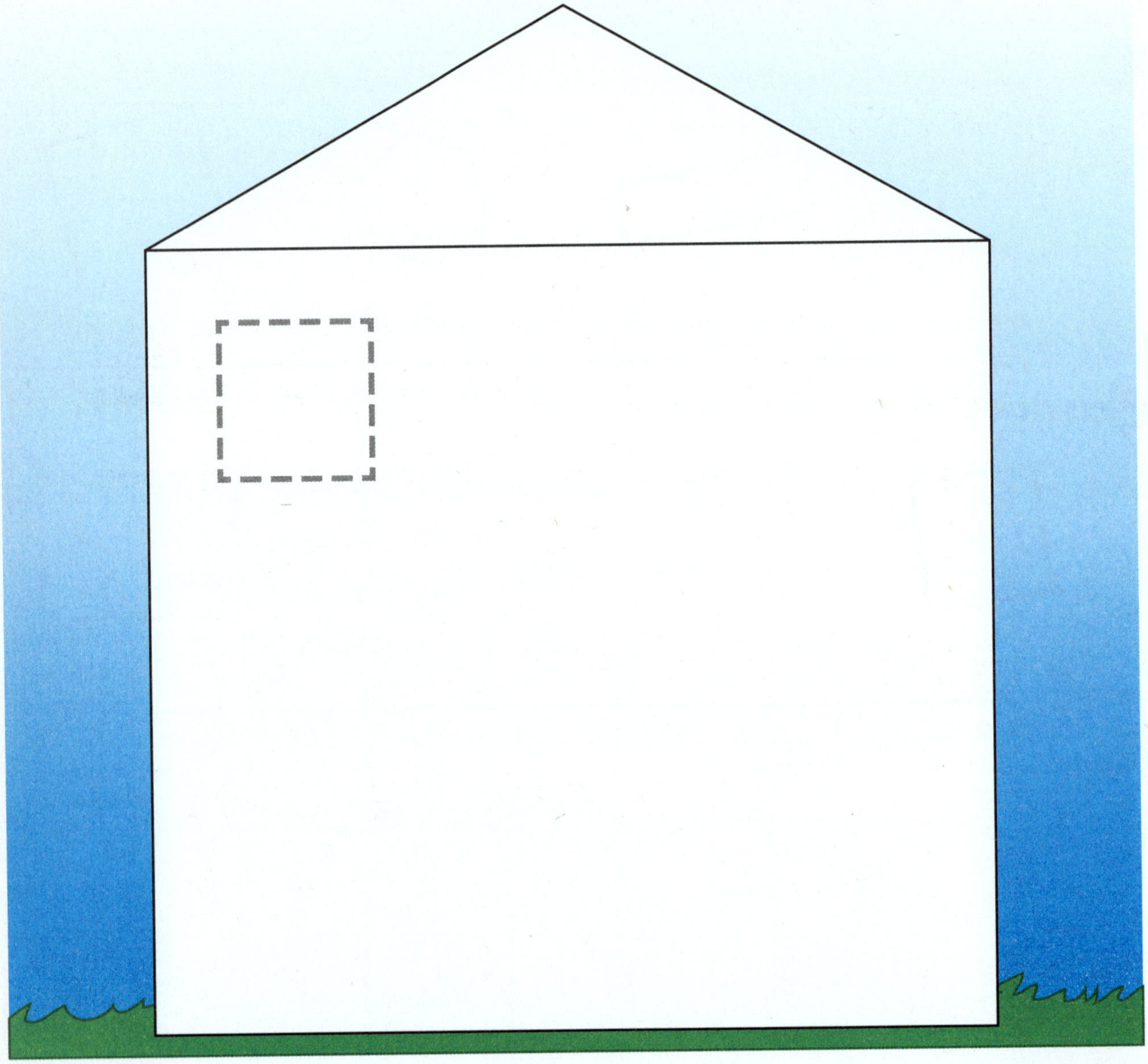

Directions:
- Use squares and rectangles to draw 6 windows and 1 door on the house. Color the house.
- Write an addition sentence to tell how many squares and how many rectangles you draw in all.

Name _____

Learning Target: Identify and describe squares.

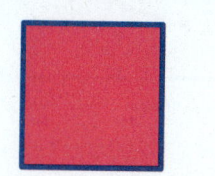 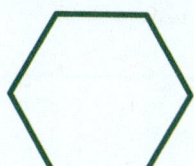  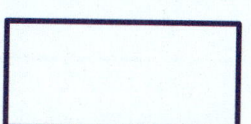

4 sides of same length, 4 L-shaped vertices

Directions: Color the square. Tell why your answer is correct.

  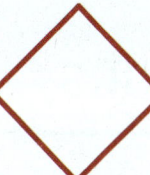 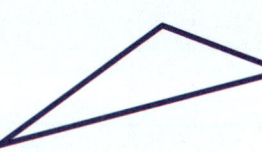

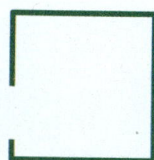  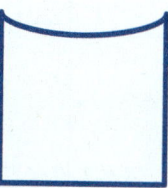 

  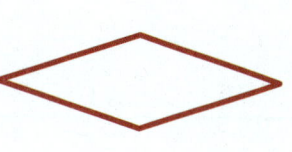 

Directions: ①—③ Color any squares. Tell why your answers are correct.

Chapter 11 | Lesson 4

4

5

_____

- - - - - - - -

_____ sides   _____ vertices

6

_____

- - - - - - - -

_____

_____

Directions: **4** Color any squares. Tell why your answers are correct. **5** Trace the shapes that are squares. Write the number of sides and the number of vertices of a square. **6** Use squares and rectangles to draw 4 windows and 1 door on the castle. Color the castle. Then write an addition sentence to tell how many squares and rectangles you draw in all.

Name _____

Learning Target: Identify and describe hexagons and circles.

 Explore and Grow

Directions:
- Use your finger to trace around the yellow hexagon. Trace and color the shapes that are hexagons.
- Use your finger to trace around the blue circle. Use a different color to trace and color the shapes that are circles.

Chapter 11 | Lesson 5

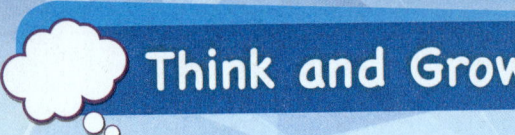

Think and Grow

6 sides,
6 vertices

0 straight sides,
0 vertices

Directions: Color any hexagons red. Color any circles blue. Tell why your answers are correct.

Apply and Grow: Practice

1

 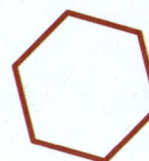  

2

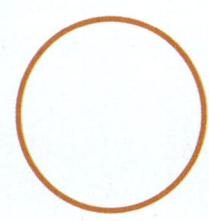   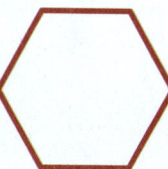

3

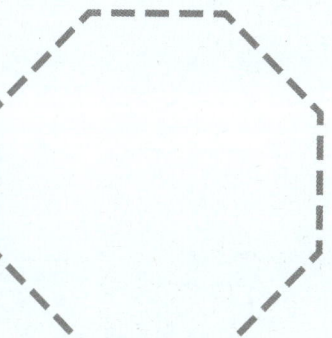 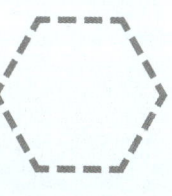 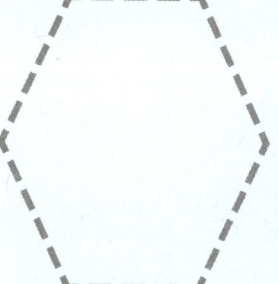 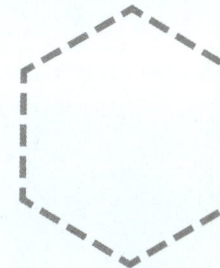

_____ _____ **sides**   _____ _____ **vertices**

Directions: **1** and **2** Color any hexagons red. Color any circles blue. Tell why your answers are correct. **3** Trace the shapes that are hexagons. Write the number of sides and the number of vertices of a hexagon.

Directions: Follow the steps to complete the robot.

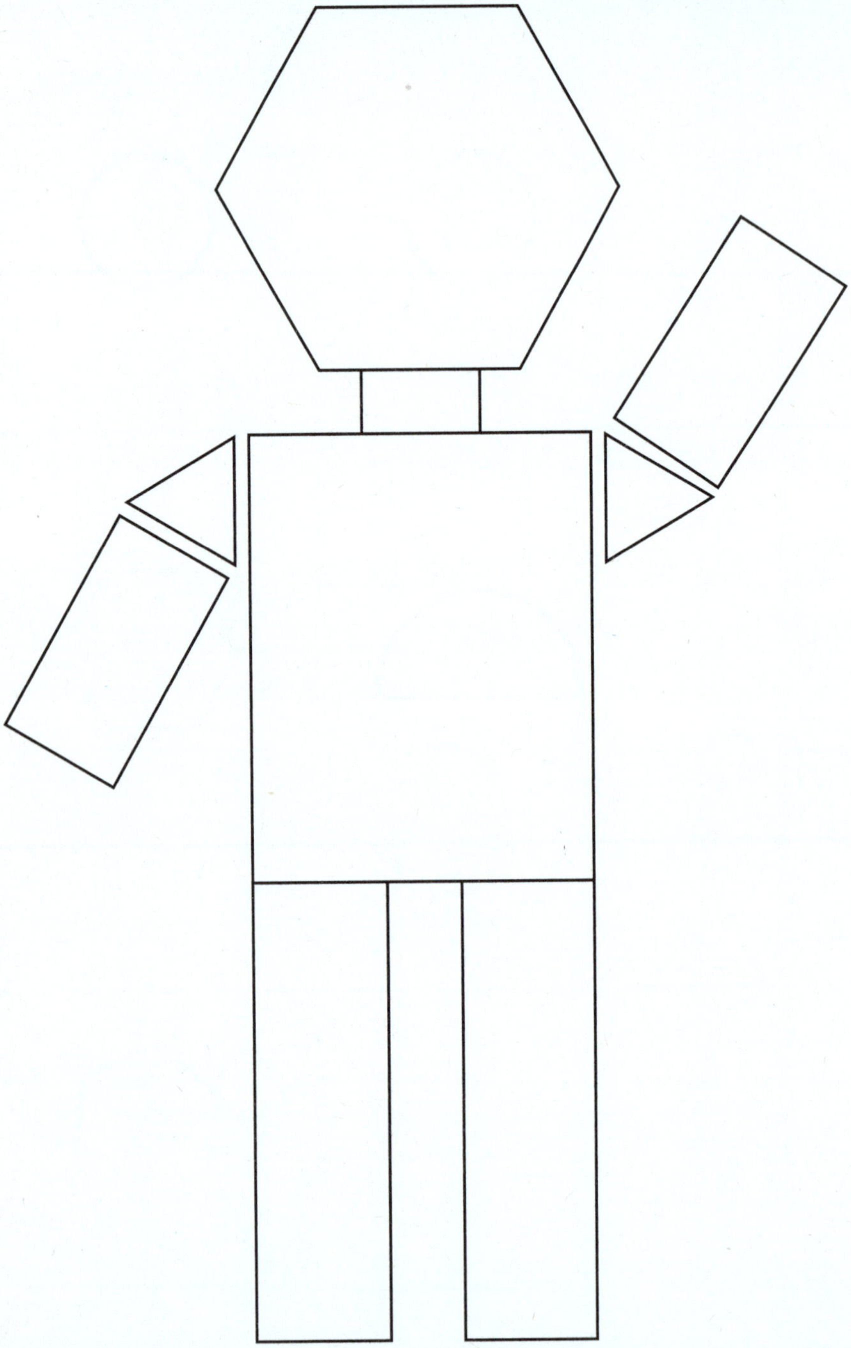

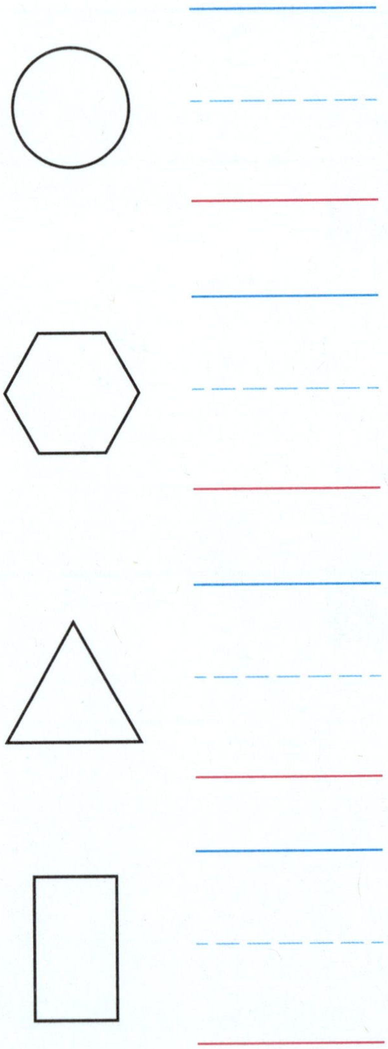

• Draw 2 circles for the robot's eyes, a triangle for the nose, and a rectangle for the mouth.

• Draw a hexagon for each of the robot's hands.

• How many of each shape is used for the whole robot? Write the number next to each shape.

• Color your robot.

574 five hundred seventy-four

Name _____

Learning Target: Identify and describe hexagons and circles.

6 sides, 6 vertices

0 straight sides, 0 vertices

Directions: Color the hexagon red. Color the circle blue. Tell why your answers are correct.

 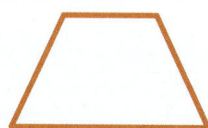 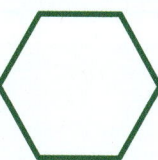  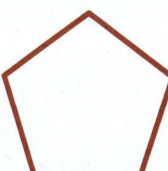

  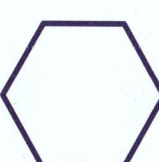  

   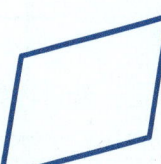 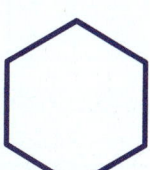

Directions: – Color any hexagons red. Color any circles blue. Tell why your answers are correct.

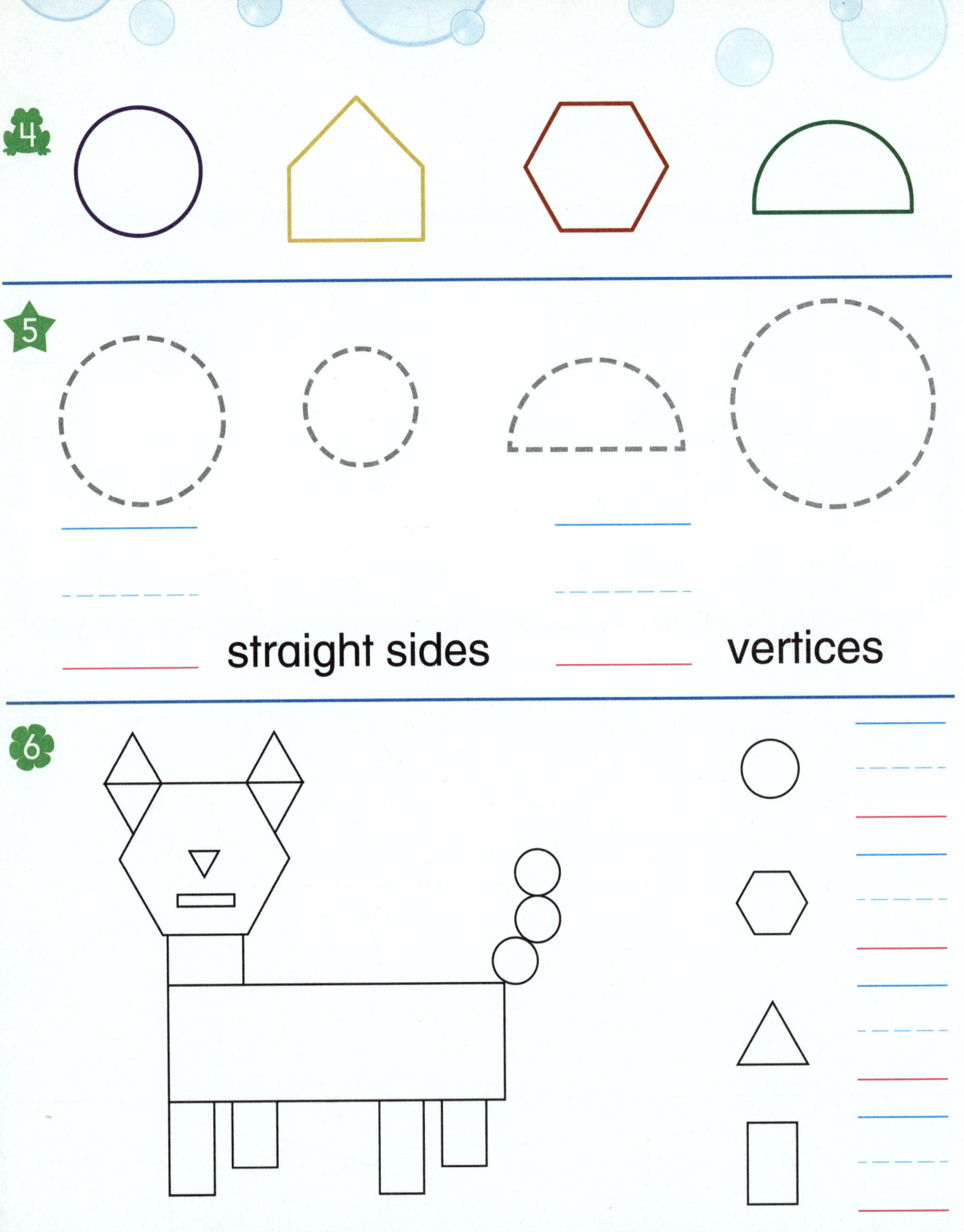

straight sides

vertices

Directions: 🐸 Color any hexagons red. Color any circles blue. Tell why your answers are correct. ⭐ Trace the shapes that are circles. Write the number of straight sides and the number of vertices of a circle. 🍀 Draw a hexagon at the end of the cat's tail. Draw 2 circles for the eyes. How many of each shape is used for the cat? Write the numbers next to the shapes. Color your cat.

Name _____

Learning Target: Join two-dimensional shapes to form a larger two-dimensional shape.

Explore and Grow

Directions:
- Use 2 squares to make a rectangle. Trace your shape.
- Add another square to make a larger rectangle. Trace your shape.
- Add another square to make a larger square. Trace your shape.

© Big Ideas Learning, LLC

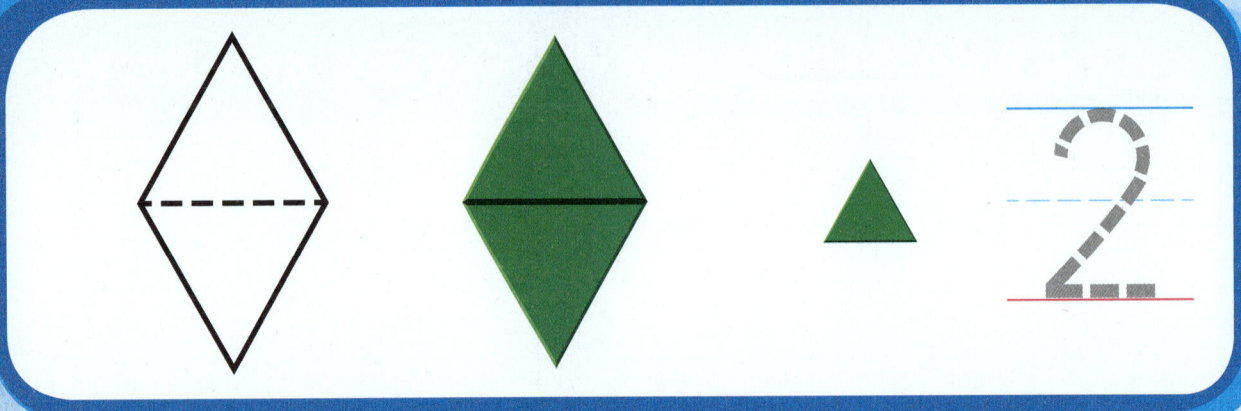

Form a larger shape.

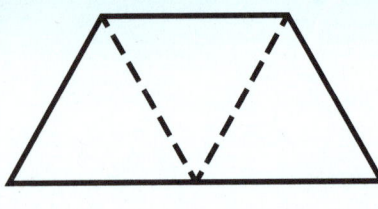  

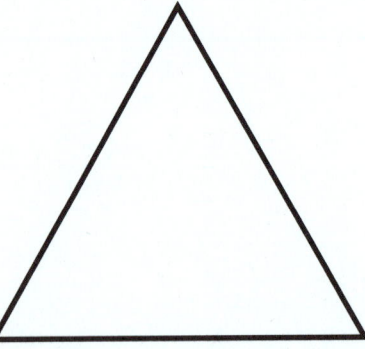

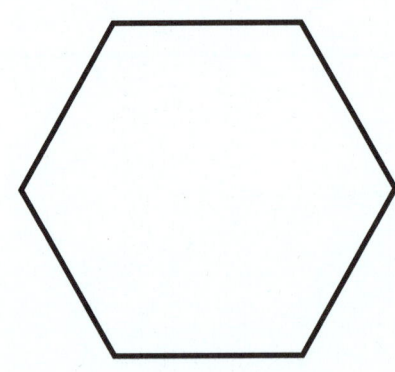

Directions: Use the pattern block shown to form the shape. Count and write how many pattern blocks you use.

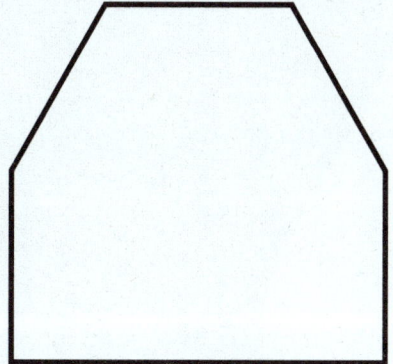

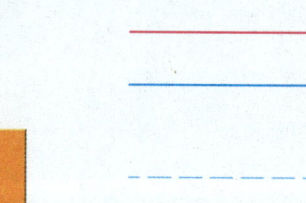

Directions: ➊ and ➋ Use the pattern blocks shown to form the shape. Count and write how many of each pattern block you use. ➌ Draw a rectangle that can be formed by the 2 triangles shown.

Think and Grow: Modeling Real Life

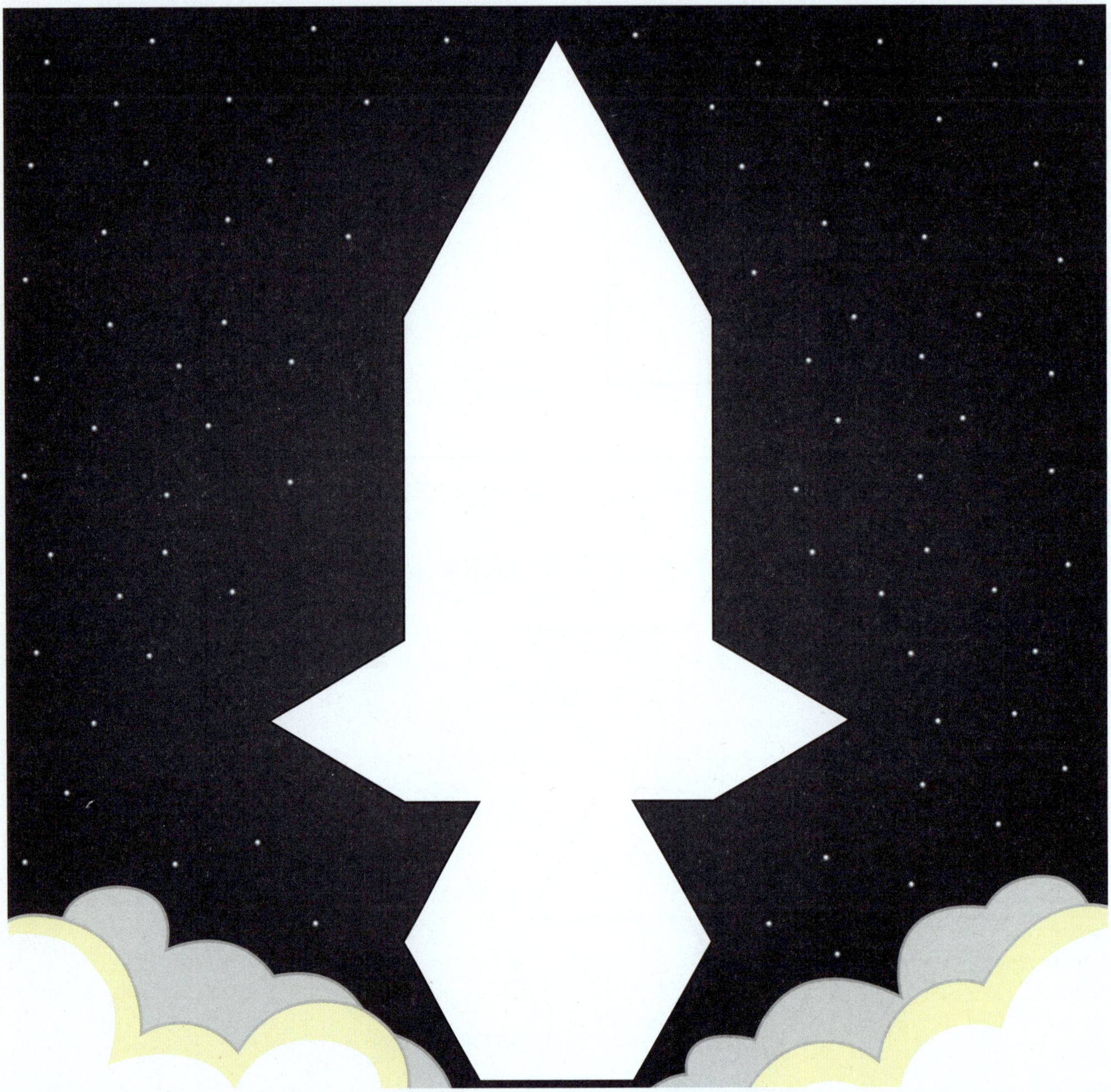

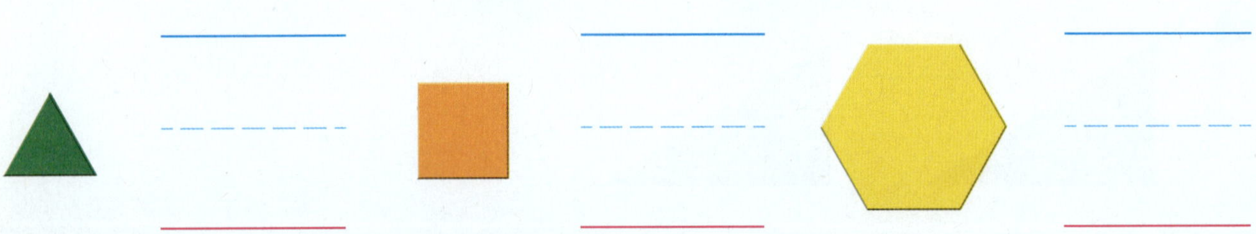

Directions: Use the pattern blocks shown to create the rocket ship. Count and write how many of each pattern block you use.

Directions: Use the pattern block shown to form the shape. Count and write how many pattern blocks you use.

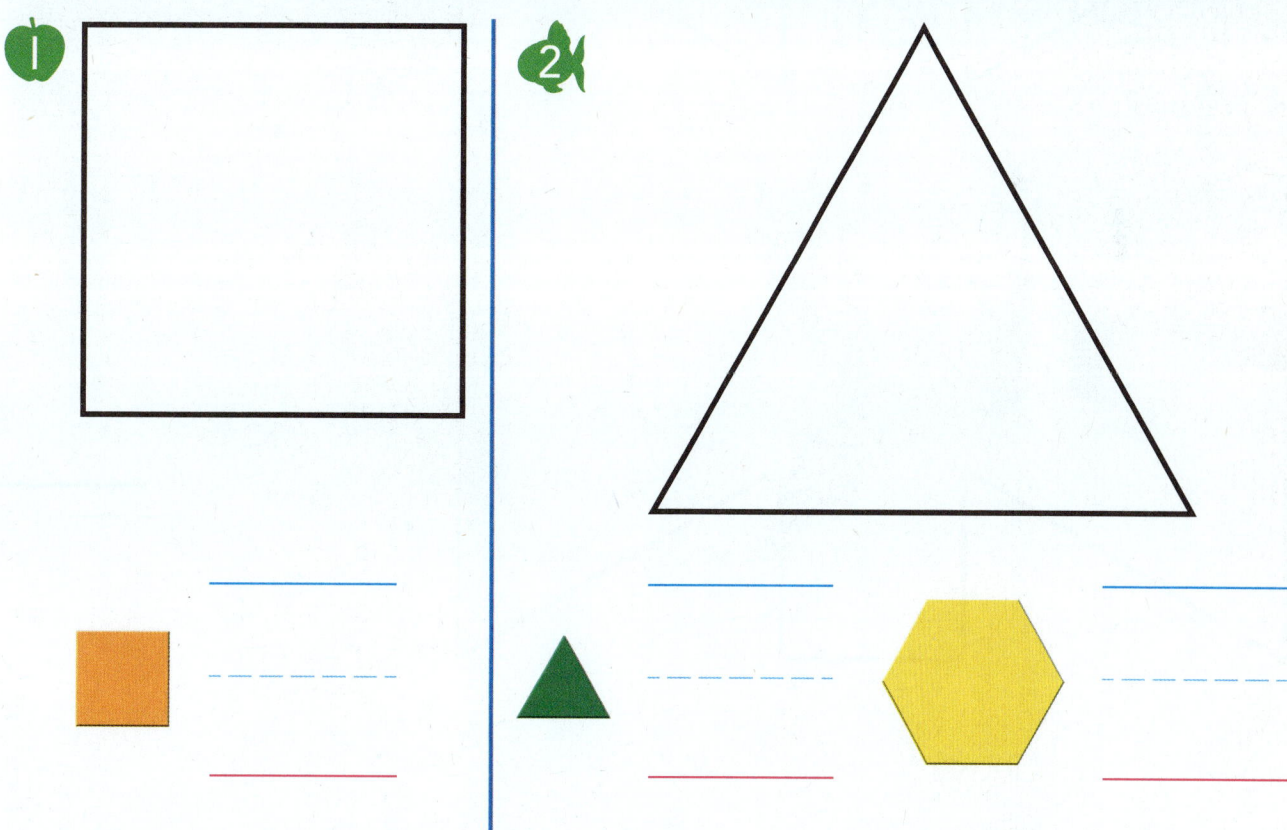

Directions: ① and ② Use the pattern blocks shown to form the shape. Count and write how many of each pattern block you use.

3

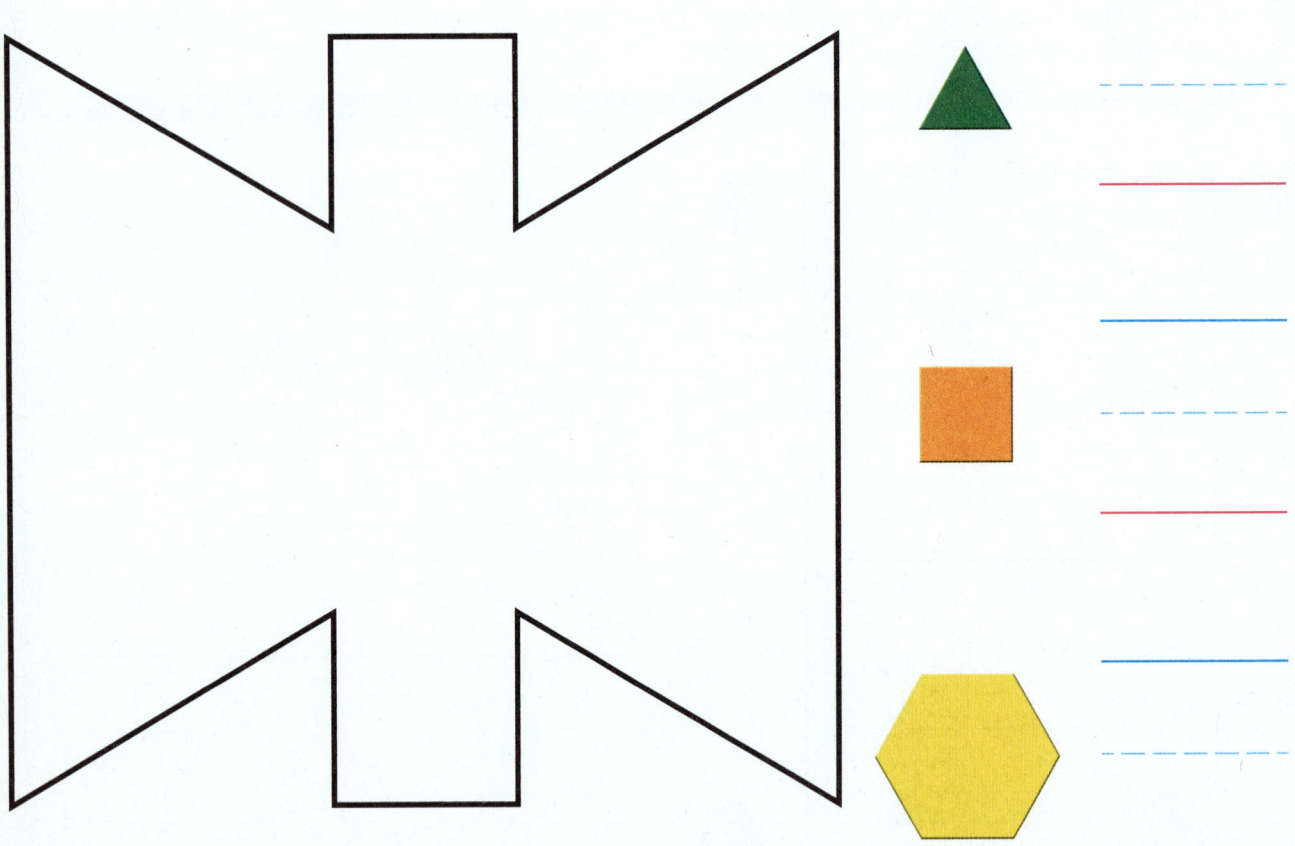

4

Directions: **3** Draw a square that can be formed by the 2 triangles shown.
4 Use the pattern blocks shown to create the butterfly. Count and write how many of each pattern block you use.

Name _____

Learning Target: Build and explore two-dimensional shapes.

 Explore and Grow

Directions: Use your materials to build one of the two-dimensional shapes shown. Circle the two-dimensional shape that you make.

Chapter 11 | **Lesson 7**

Directions:

- Use your materials to build a triangle. Draw your triangle or attach it to the page.
- Use your materials to build a rectangle. Draw your shape or attach it to the page.

Name _____

Directions: Use your materials to build a hexagon. Draw your hexagon or attach it to the page. Use your materials to build a circle. Draw your circle or attach it to the page. Use your materials to build a two-dimensional shape that has 4 vertices. Then build a different shape that has 4 vertices. Draw your shapes or attach them to the page.

Think and Grow: Modeling Real Life

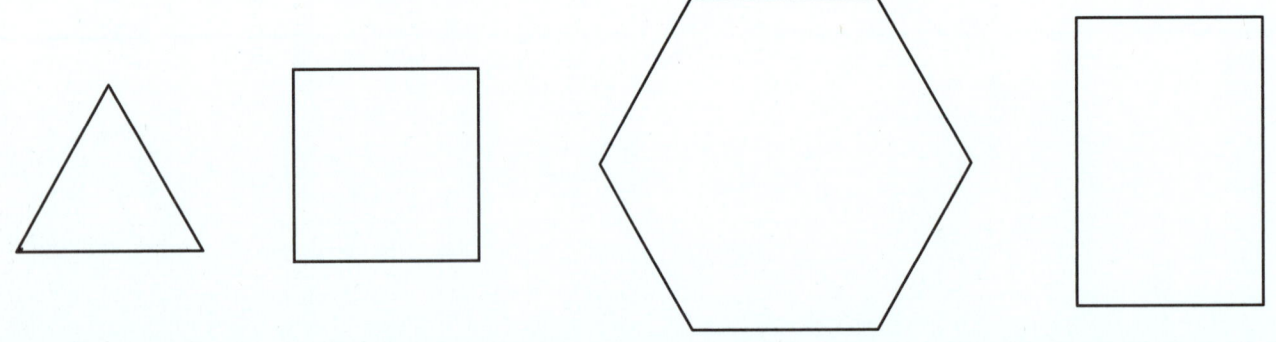

Directions:

- Use your materials to build the front of the house in the picture. Draw your shape or attach it to the page.
- Circle the shapes that you use to make the front of the house.

Name _____

Learning Target: Build and explore two-dimensional shapes.

4 sides of equal length,
4 L-shaped vertices

Directions: Use your materials to build a square. Draw your square or attach it to the page.

Directions: 1 – 3 Use your materials to build the shape shown. Draw your shape or attach it to the page. 4 Use your materials to build a two-dimensional shape that has 6 vertices. Draw your shape or attach it to the page.

Chapter 11 | Lesson 7

⑤

---

❻

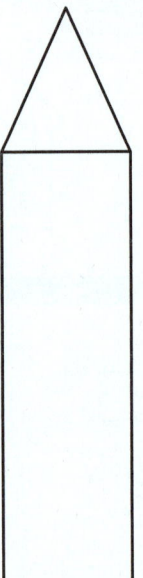

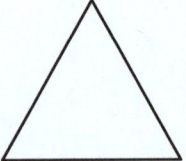 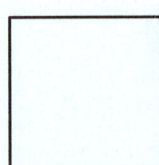 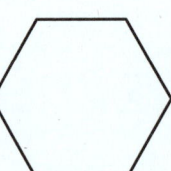 

Directions: **⑤** Use your materials to build a two-dimensional shape that is not a rectangle. Then build a different two-dimensional shape that is not a rectangle. Draw your shapes or attach them to the page. **❻** Use your materials to build the front of the sand castle tower in the picture. Draw your shape or attach it to the page. Circle the shapes that you use to make the front of the tower.

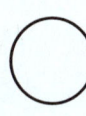 _____ 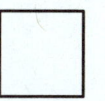 _____ 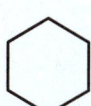 _____

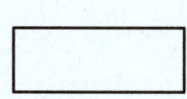 _____  _____

Directions: ❶ Use the clues to draw each two-dimensional shape to make an animal.

- The face is a shape that has 1 more than 5 sides.
- The eyes are shapes that are curved and have no vertices.
- The nose and ears are shapes that have 1 more than 2 vertices.
- The mouth, the body, and the tail are shapes that have more than 3 sides, but less than 5 sides and have L-shaped vertices.
- The 4 legs are shapes that have 1 less than 5 sides and have all equal side lengths.

❷ Count and write how many of each shape you draw.

Shape Flip and Find

Directions: Place the Shape Flip and Find Cards facedown in the boxes. Take turns flipping 2 cards. If your cards show the same shape, keep the cards. If your cards show different shapes, flip the cards back over. Repeat until all cards have been used.

Describe Two-Dimensional Shapes

1

 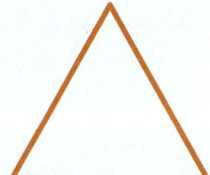  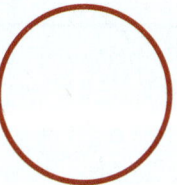

2

 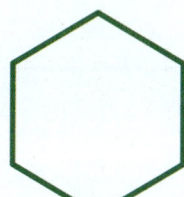  

3

 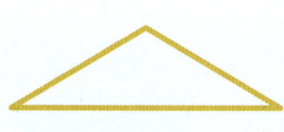  

4

Directions: **1** Color the shape that has only 3 vertices. **2** Color the shape that has a curve. **3** Color the shapes that have 6 sides. **4** Color the shapes that have 4 vertices.

11.2 Triangles

 5

 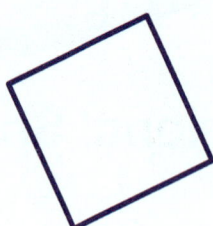  

 6

  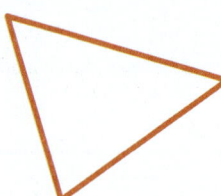 

11.3 Rectangles

 7

 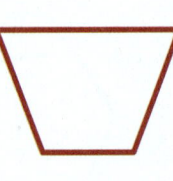  

8

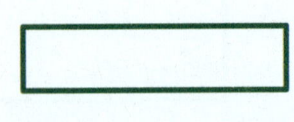 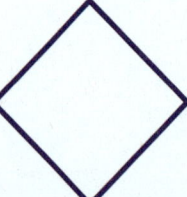 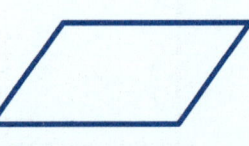 

Directions: **5** and **6** Color the triangle. Tell why your answer is correct.
7 and **8** Color any rectangles. Tell why your answers are correct.

11.4 Squares

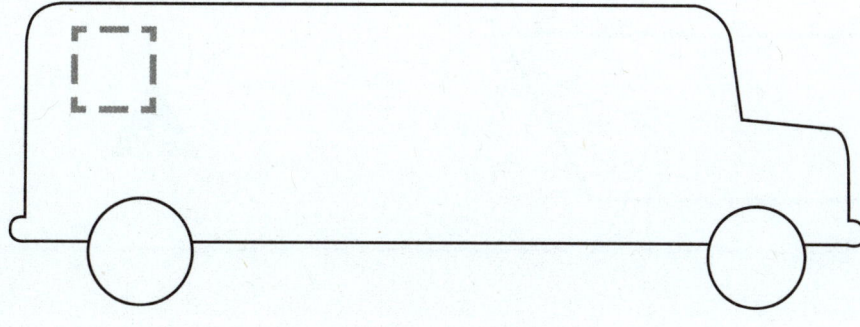

11.5 Hexagons and Circles

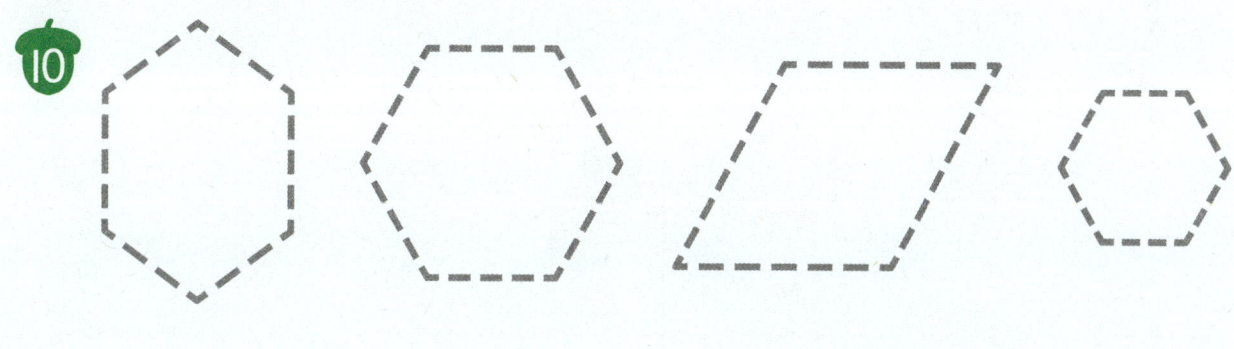

_____ sides   _____ vertices

Directions: Use squares and rectangles to draw 5 windows and 1 door on the bus. Color the bus. Then write an addition sentence to tell how many squares and rectangles you draw in all. Trace the shapes that are hexagons. Write the number of sides and the number of vertices of a hexagon.

11.6 Join Two-Dimensional Shapes

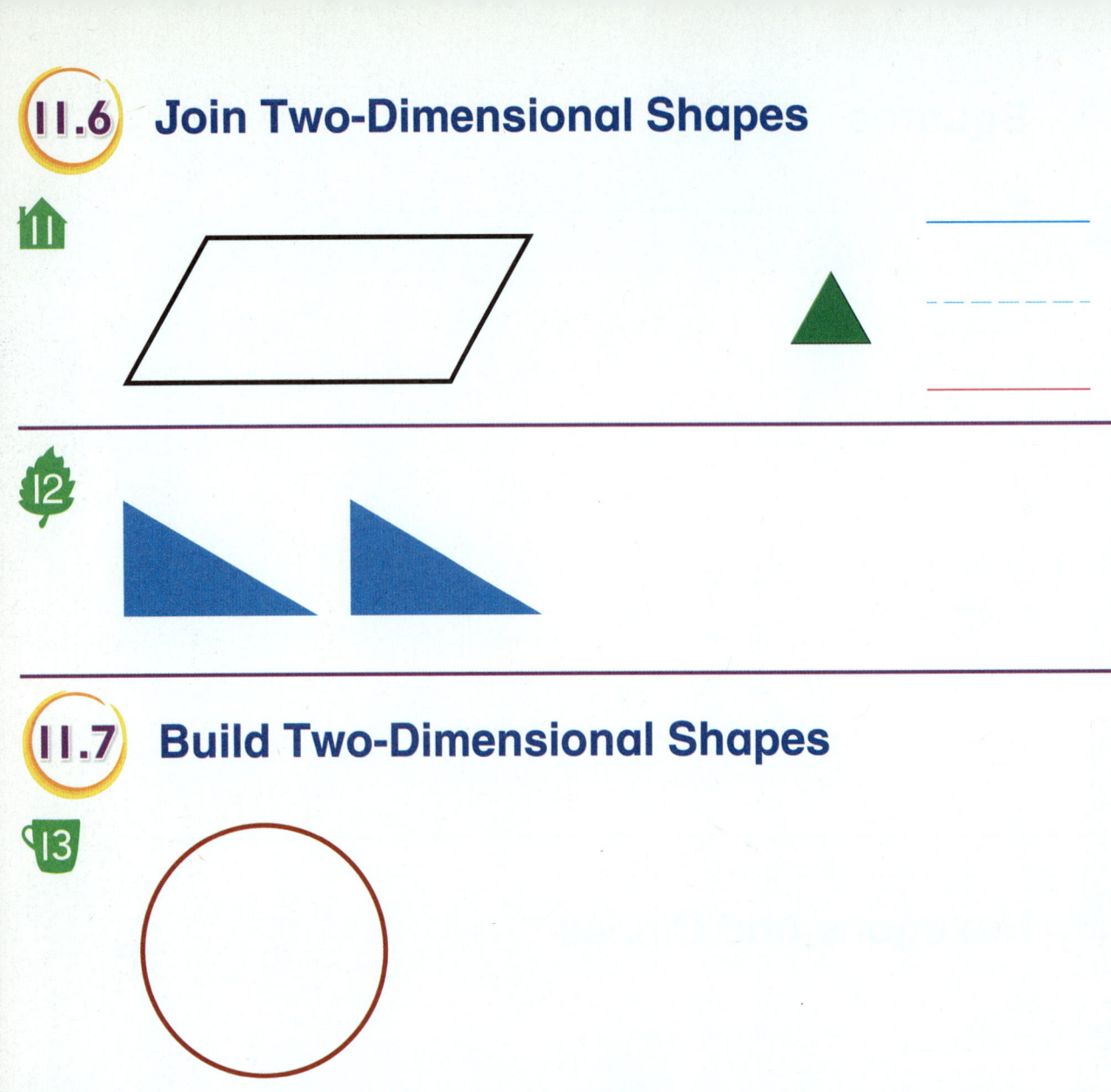

11.7 Build Two-Dimensional Shapes

Directions: 🏠 Use the pattern block shown to form the shape. Count and write how many pattern blocks you use. 🍃 Draw a larger triangle that can be formed by the 2 triangles shown. 🍵 Use your materials to build the shape. Draw your shape or attach it to the page. 👕 Use your materials to build a two-dimensional shape that has 4 sides of the same length. Draw your shape or attach it to the page. 🦋 Use your materials to build a two-dimensional shape that is *not* a hexagon. Draw your shape or attach it to the page.

- **What items do you recycle?**
- **What three-dimensional shapes do you see in the picture?**

Chapter Learning Target:
Understand three-dimensional shapes.

Chapter Success Criteria:
- ■ I can identify three-dimensional shapes.
- ■ I can describe three-dimensional shapes.
- ■ I can compare three-dimensional shapes.
- ■ I can build three-dimensional shapes.

Vocabulary

Review Words
circle
square
two-dimensional shape

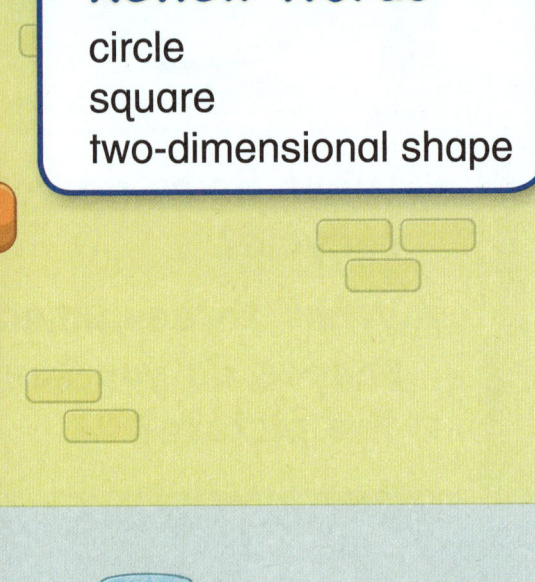

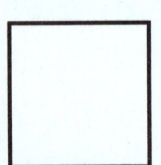

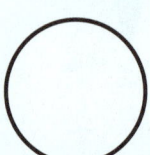

Directions: Circle each can. Draw a square around each box. Count and write how many of each two-dimensional shape you draw.

Chapter 12 Vocabulary Cards

above

behind

below

beside

cone

cube

curved surface

cylinder

© Big Ideas Learning, LLC

© Big Ideas Learning, LLC

© Big Ideas Learning, LLC

© Big Ideas Learning, LLC

© Big Ideas Learning, LLC

© Big Ideas Learning, LLC

© Big Ideas Learning, LLC

© Big Ideas Learning, LLC

Chapter 12 Vocabulary Cards

flat surface

in front of

next to

roll

slide

sphere

stack

three-dimensional shape

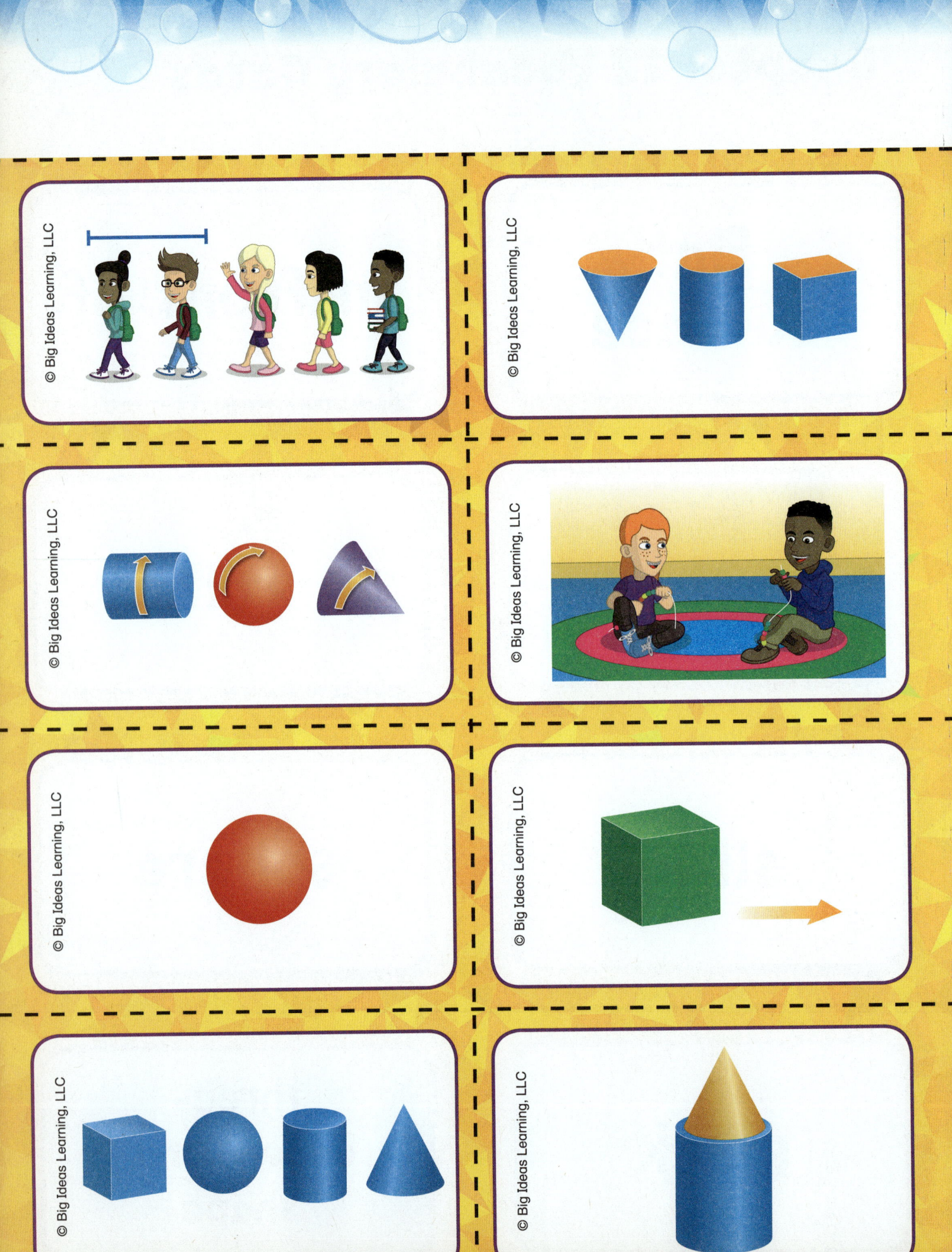

© Big Ideas Learning, LLC

© Big Ideas Learning, LLC

© Big Ideas Learning, LLC

© Big Ideas Learning, LLC

© Big Ideas Learning, LLC

© Big Ideas Learning, LLC

© Big Ideas Learning, LLC

© Big Ideas Learning, LLC

vertex

vertices

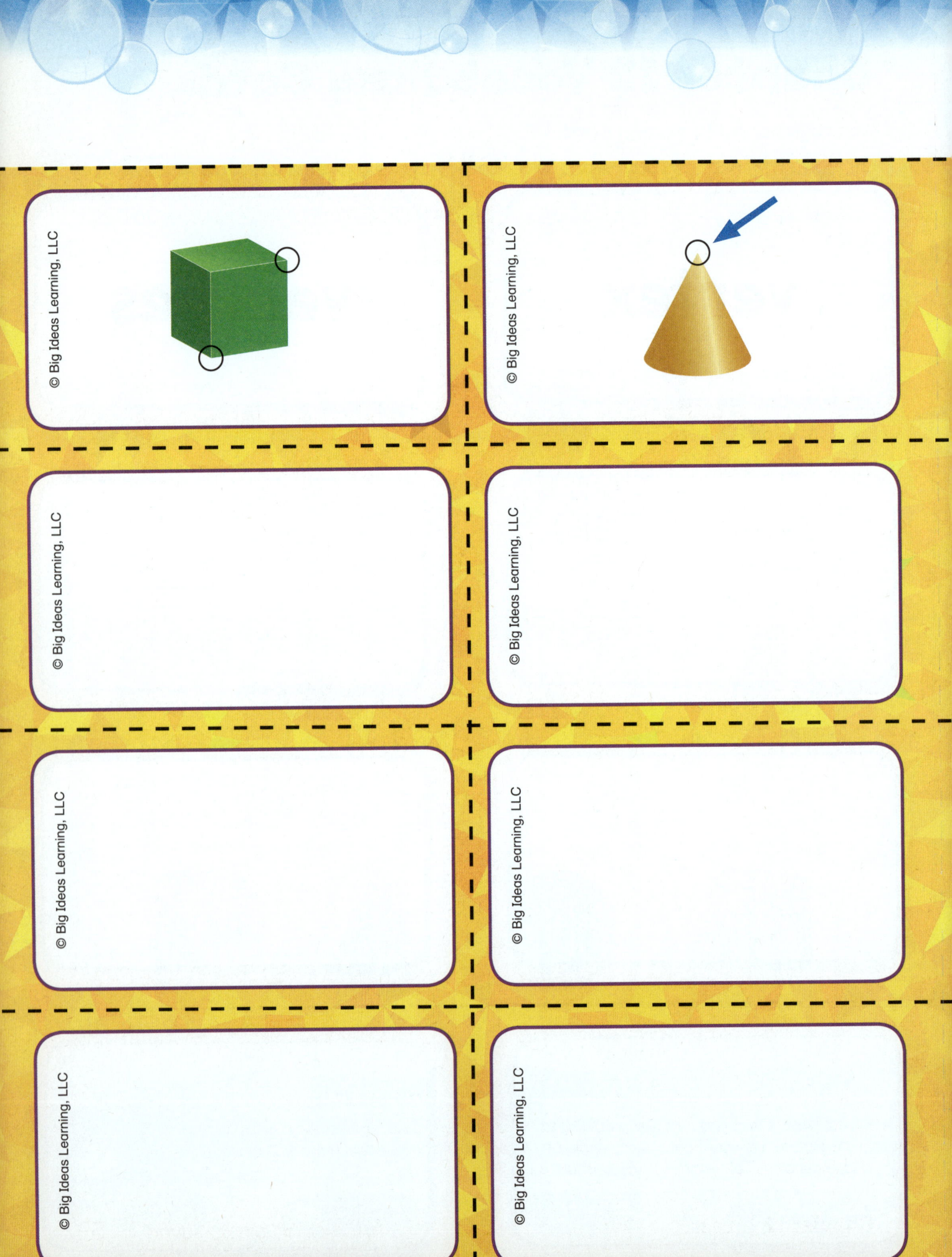

Learning Target: Identify
and describe two-dimensional and
three-dimensional shapes.

Explore and Grow

Directions: Circle any triangles, rectangles, squares, hexagons, and circles you see
in the picture. Use another color to circle any objects in the picture that match the blue
shapes shown. Tell what you notice about each shape.

Think and Grow

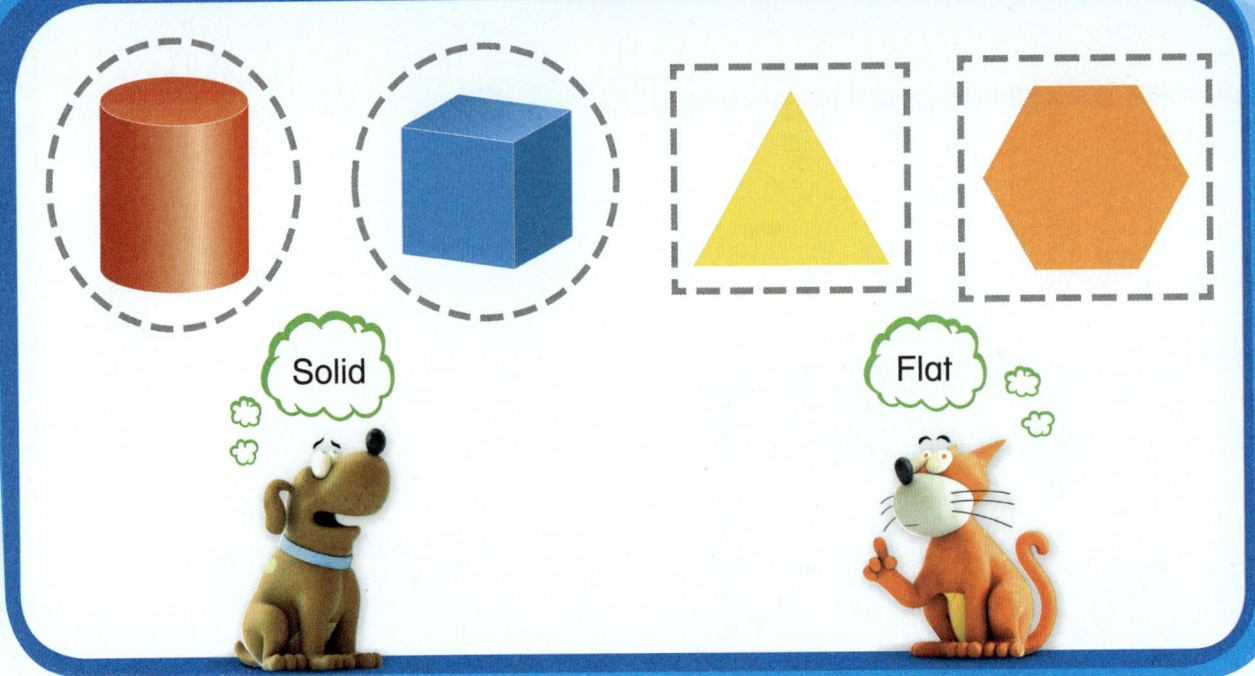

Solid

Flat

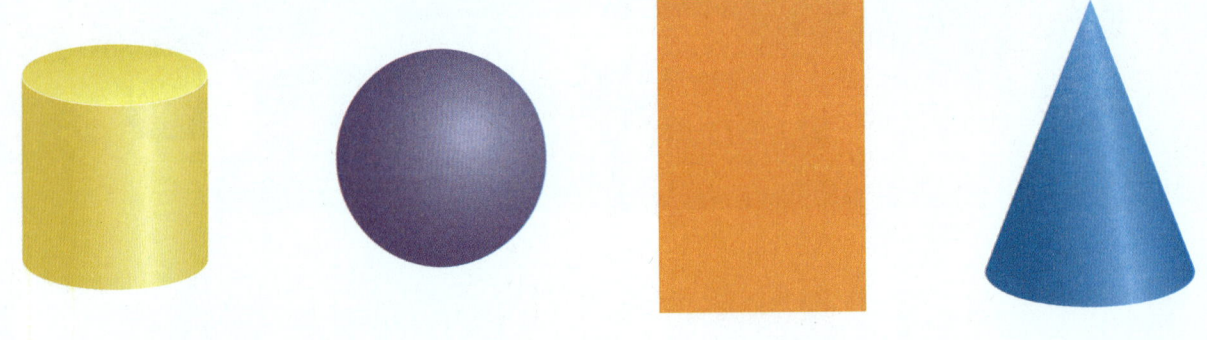

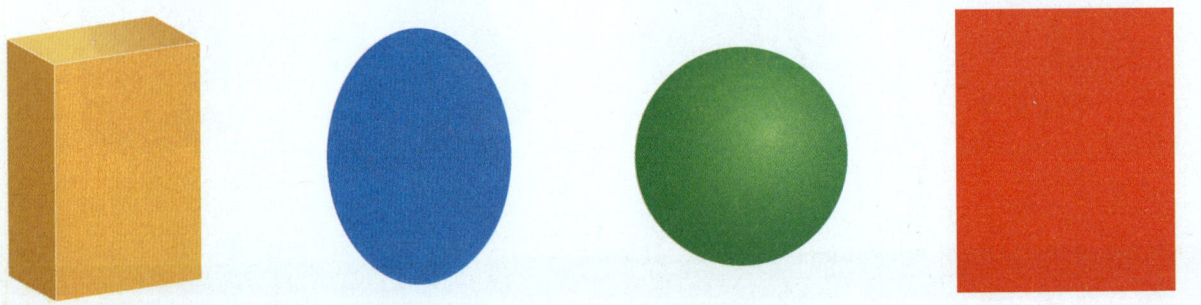

Directions: Circle any three-dimensional shapes. Draw rectangles around any two-dimensional shapes. Tell why your answers are correct.

Name _____

 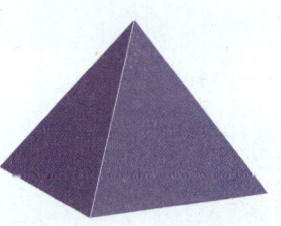  

 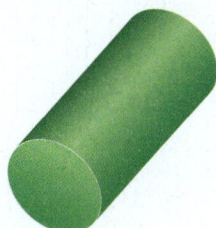  

Directions: ①–④ Circle any three-dimensional shapes. Draw rectangles around any two-dimensional shapes. Tell why your answers are correct.

Think and Grow: Modeling Real Life

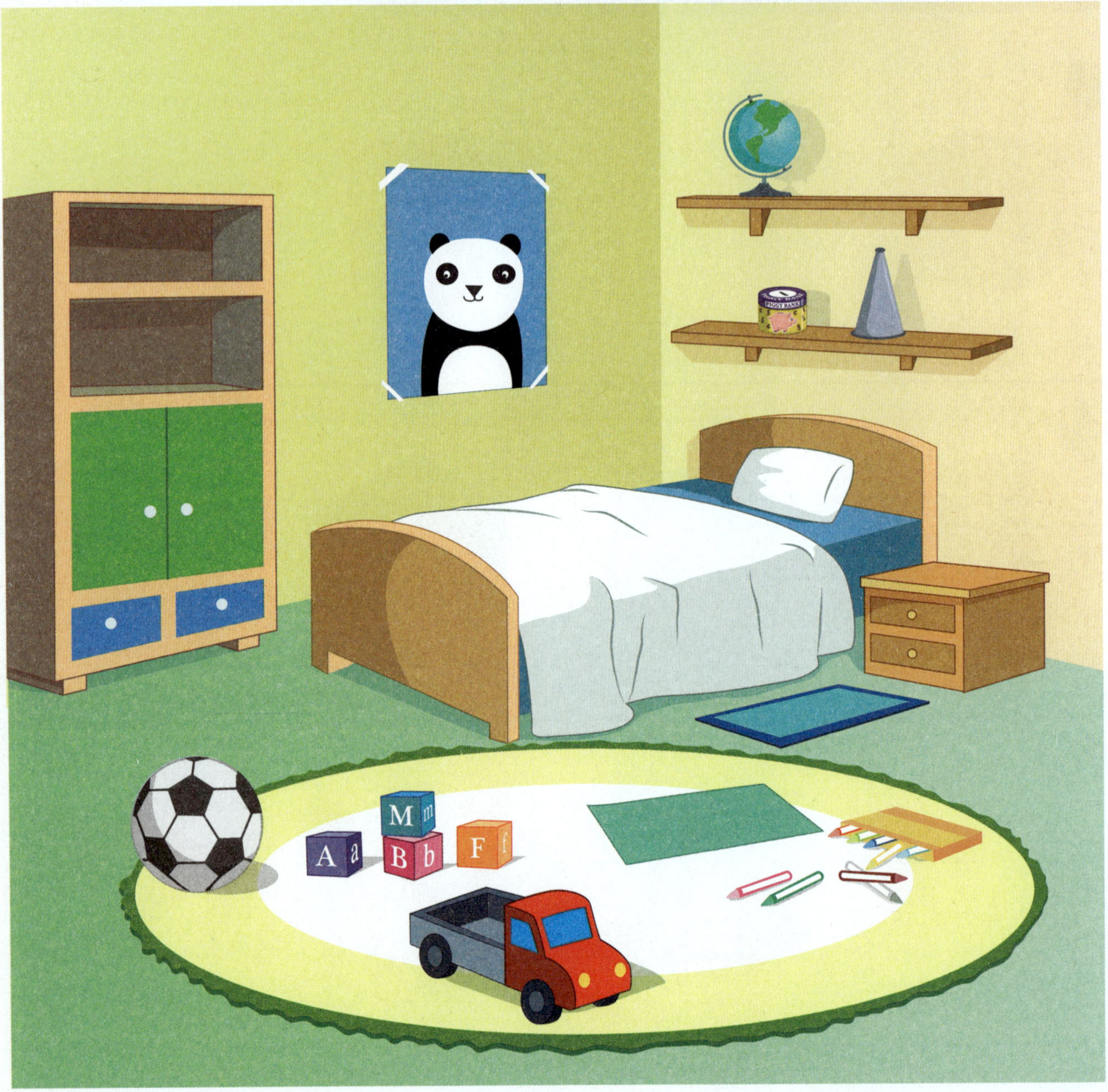

three-dimensional

two-dimensional

Directions: Circle any shapes in the picture that are solids. Draw rectangles around any shapes in the picture that are flats. Count and write how many solids and flats you find.

Name ______________________

Learning Target: Identify and describe two-dimensional and three-dimensional shapes.

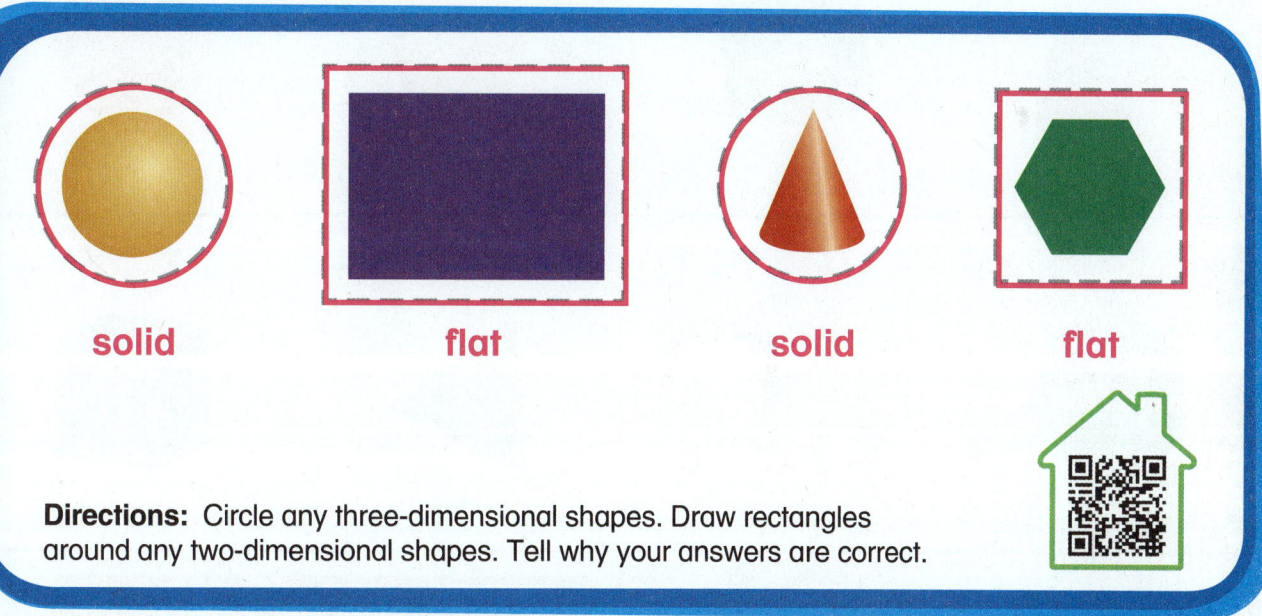

solid
flat
solid
flat

Directions: Circle any three-dimensional shapes. Draw rectangles around any two-dimensional shapes. Tell why your answers are correct.

1

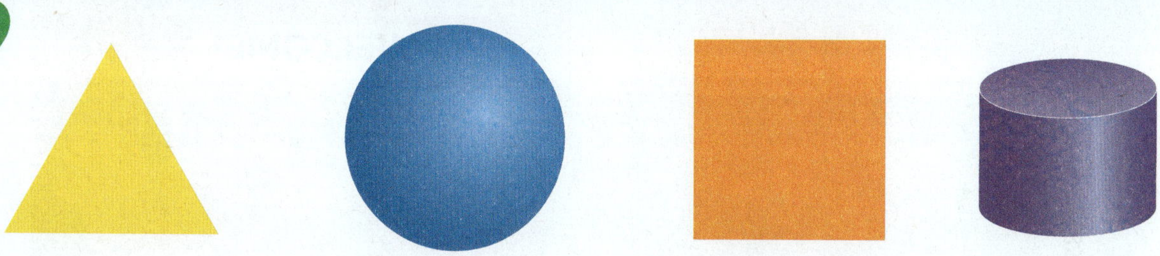

2

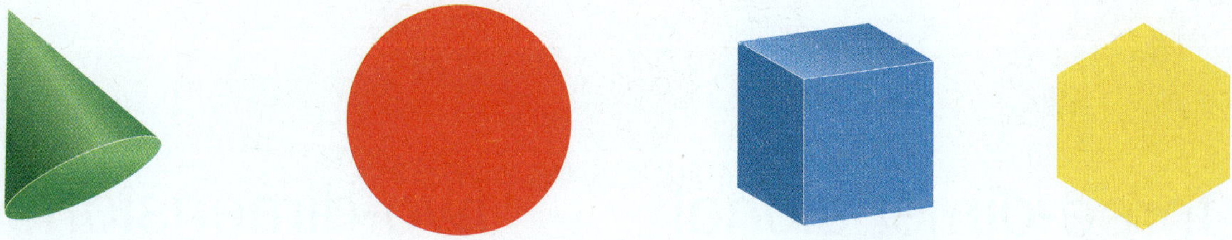

3

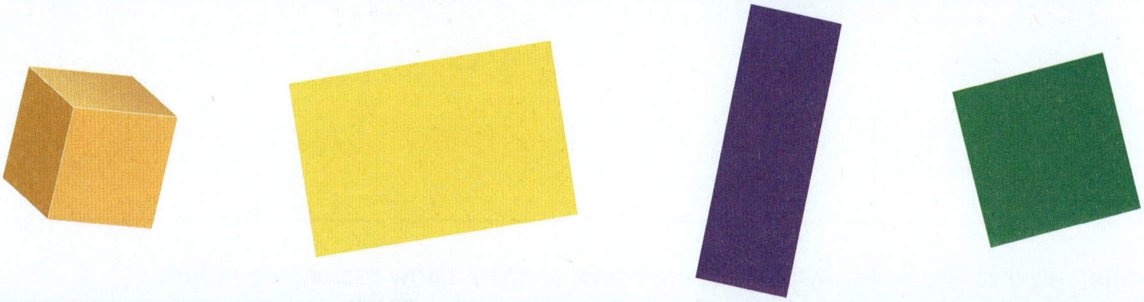

Directions: 1–3 Circle any three-dimensional shapes. Draw rectangles around any two-dimensional shapes. Tell why your answers are correct.

three-dimensional   two-dimensional

_____   _____

- - - - - - - - - - -   - - - - - - - - - - -

_____   _____

© Big Ideas Learning, LLC

Directions: and Circle any three-dimensional shapes. Draw rectangles around any two-dimensional shapes. Tell why your answers are correct. Circle any three-dimensional shapes in the picture. Count and write the number. Draw rectangles around any two-dimensional shapes in the picture. Count and write the number.

Learning Target: Describe three-dimensional shapes.

 **Explore and Grow**

rolls

stacks

slides

Directions: Cut out the Roll, Stack, Slide Sort Cards. Sort the cards into the categories shown.

 **Think and Grow**

Roll it!

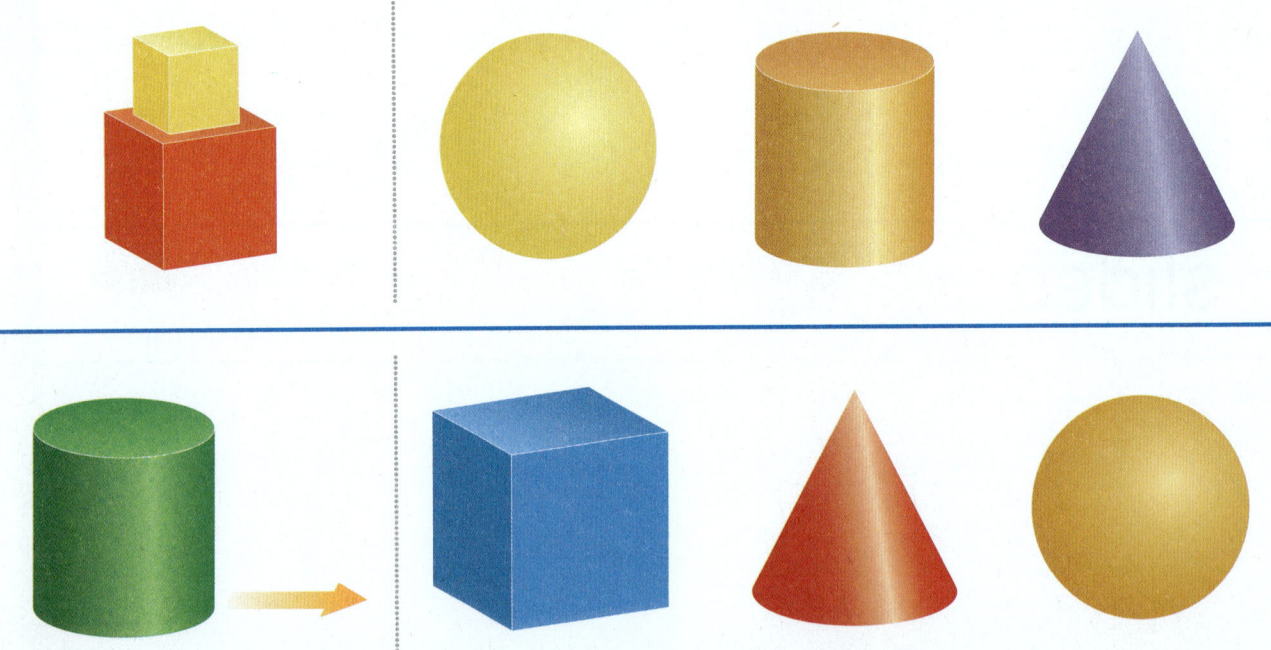

Directions:
- Look at the solid shape on the left that rolls. Circle the other solid shapes that roll.
- Look at the solid shapes on the left that stack. Circle the other solid shapes that stack.
- Look at the solid shape on the left that slides. Circle the other solid shapes that slide.

604   six hundred four

✔ Apply and Grow: Practice

   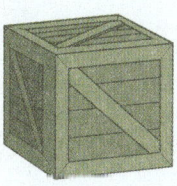 

Directions: ❶ Look at the solid shape on the left that rolls. Circle the other solid shapes that roll. ❷ Circle the solid shapes that roll and slide. ❸ Circle the solid shapes that stack and slide. ❹ Circle the solid shape that does *not* stack or slide.

_____

- - - - - - - - -

_____

_____

- - - - - - - - -

_____

Directions: You stack the 3 objects shown. Write 1 below the object you place at the bottom of the stack, write 2 below the object you stack next, and write 3 below the object you stack last. Tell why you chose this order.

Name ______________________________

Learning Target: Describe three-dimensional shapes.

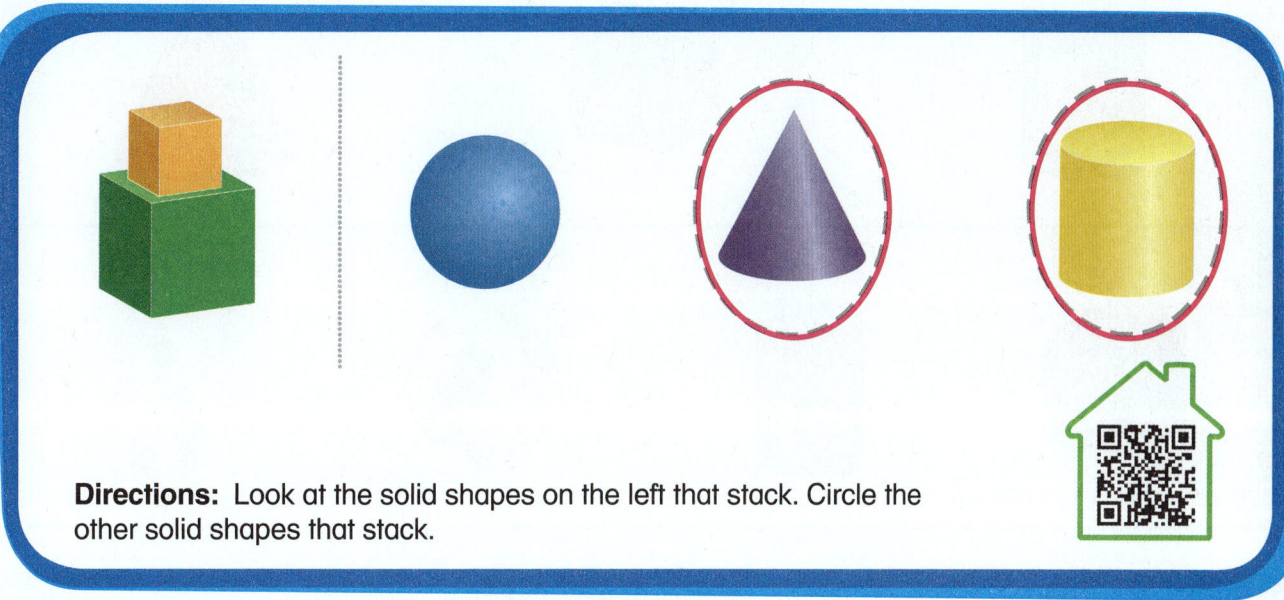

Directions: Look at the solid shapes on the left that stack. Circle the other solid shapes that stack.

1

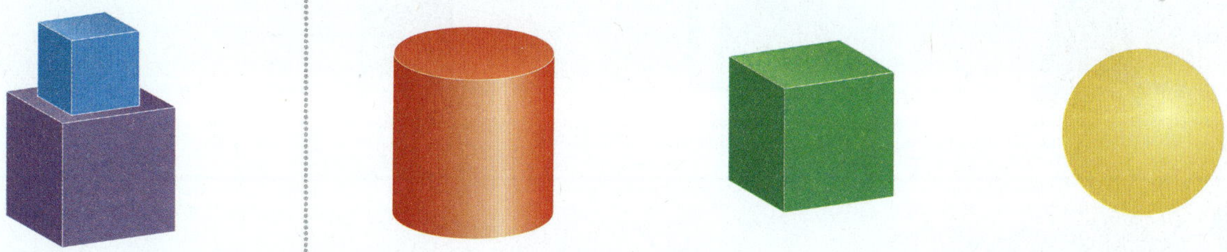

2

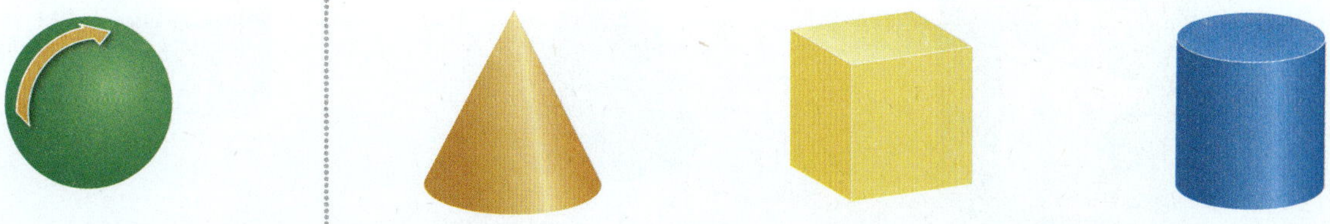

3

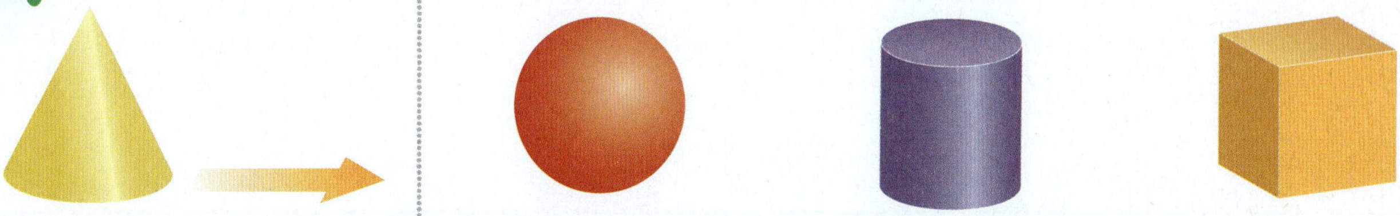

Directions: 1 Look at the solid shapes on the left that stack. Circle the other solid shapes that stack. 2 Look at the solid shape on the left that rolls. Circle the other solid shapes that roll. 3 Look at the solid shape on the left that slides. Circle the other solid shapes that slide.

4

 5

 6

 7

_____   _____   _____

------   ------   ------

_____   _____   _____

Directions: **4** Circle the solid shapes that roll and stack. **5** Circle the solid shapes that stack and slide. **6** Circle the solid shape that does *not* roll. **7** You stack the 3 objects shown. Write 1 below the object you place at the bottom of the stack, write 2 below the object you stack next, and write 3 below the object you stack last. Tell why you chose this order.

Name _____

Learning Target: Identify and describe cubes and spheres.

 **Explore and Grow**

cube

sphere

Directions: Cut out the Cube and Sphere Sort Cards. Sort the cards into the categories shown.

Think and Grow

Square, flat surfaces

Curved surface, no flat surfaces

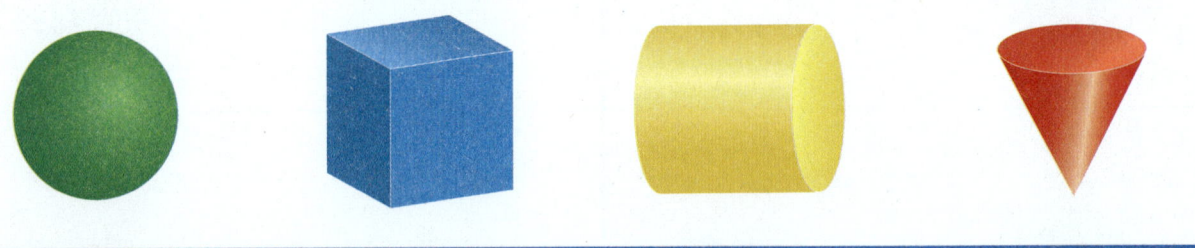

Directions: Circle the cube. Draw a rectangle around the sphere. Tell why your answers are correct.

Apply and Grow: Practice

 1

 2

 3

 4

Directions: **1** Circle the cube. Draw a rectangle around the sphere. Tell why your answers are correct. **2**–**4** Circle any object that looks like a cube. Draw a rectangle around any object that looks like a sphere. Tell why your answers are correct.

_____

_ _ _ _ _ _ _

_____ flat surfaces

Directions: Use Make a Cube to build your own number cube. Draw the shape of the flat surfaces of your cube. Count and write the number of flat surfaces.

612　six hundred twelve

Learning Target: Identify and describe cubes and spheres.

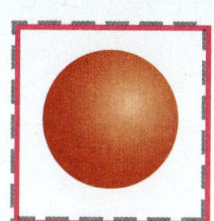

curved surface,
no flat surfaces

square,
flat surfaces

Directions: Circle the cube. Draw a rectangle around the sphere.
Tell why your answers are correct.

Directions: ❶–❸ Circle the cube. Draw a rectangle around the sphere. Tell why
your answers are correct.

Chapter 12 | **Lesson 3**

_____

_ _ _ _ _ _ _

_____ flat surfaces

© Big Ideas Learning, LLC

Directions: **4**–**6** Circle any object that looks like a cube. Draw a rectangle around any object that looks like a sphere. Tell why your answers are correct. **7** Draw the shape of the flat surfaces of a die. Count and write the number of flat surfaces.

Name _____

Learning Target: Identify and describe cones and cylinders.

 **Explore and Grow**

cone

cylinder

Directions: Cut out the Cone and Cylinder Sort Cards. Sort the cards into the categories shown.

Think and Grow

Curved surface, flat surface

Curved surface, flat surfaces

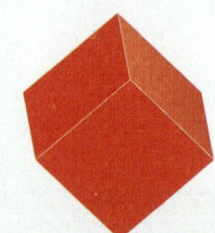   

Directions: Circle the cone. Draw a rectangle around the cylinder. Tell why your answers are correct.

 **Apply and Grow: Practice**

 1

 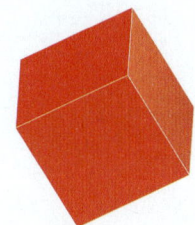  

 2

 3

 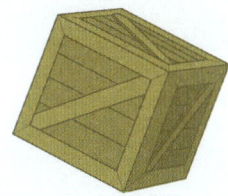  

4

Directions: **1** Circle the cone. Draw a rectangle around the cylinder. Tell why your answers are correct. **2**–**4** Circle any object that looks like a cone. Draw a rectangle around any object that looks like a cylinder. Tell why your answers are correct.

_____

- - - - - - -

_____ **flat surfaces**

Directions: Use Make a Cylinder to build a can of vegetables. Draw the shape of the flat surfaces of your can. Count and write the number of flat surfaces.

Name _____

Learning Target: Identify and describe cones and cylinders.

curved surface, flat surfaces   curved surface, flat surface

Directions: Circle the cone. Draw a rectangle around the cylinder. Tell why your answers are correct.

 1

 2

   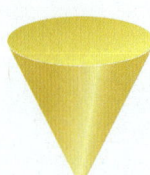

 3

Directions: **1**–**3** Circle the cone. Draw a rectangle around the cylinder. Tell why your answers are correct.

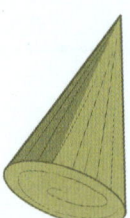

_____

_ _ _ _ _

_____ flat surface

Directions: – Circle any object that looks like a cone. Draw a rectangle around any object that looks like a cylinder. Tell why your answers are correct. Draw the shape of the flat surface of a cone. Count and write the number of flat surfaces.

Name _____

Learning Target: Build and explore three-dimensional shapes.

 **Explore and Grow**

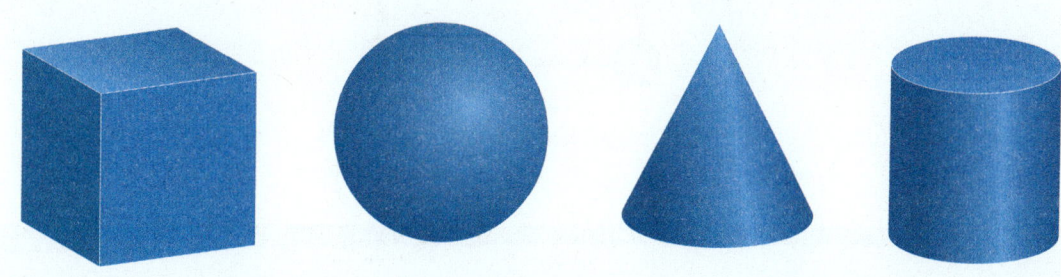

Directions: Use your materials to build one of the three-dimensional shapes shown. Circle the three-dimensional shape that you build.

Think and Grow

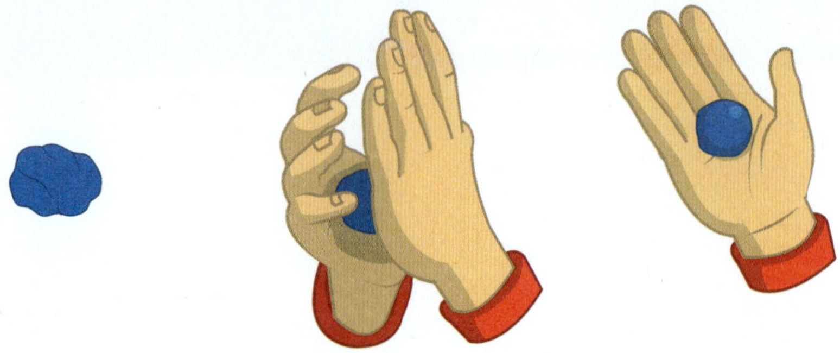

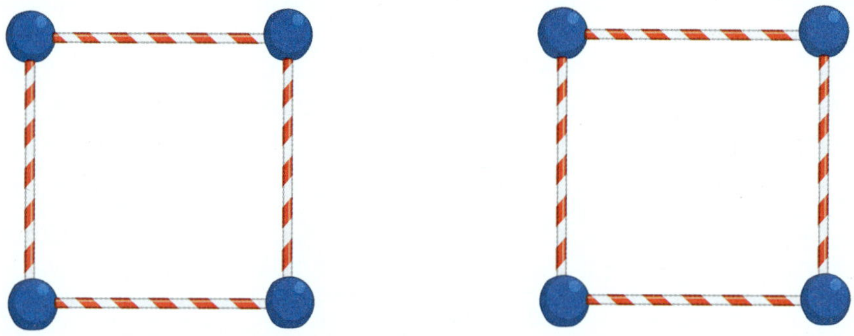

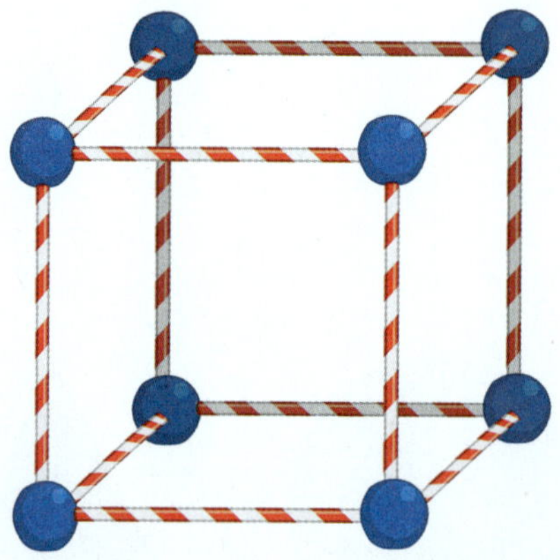

Directions:

- Use your materials to build the 2 shapes shown.
- Connect the 2 shapes that you build, as shown.
- Tell what solid shape you build.

Name _____

Directions: – Use your materials to build the solid shape shown. Use your materials to build a solid shape that has 6 square, flat surfaces. Circle the shape you build.

Directions:
- Use your materials to build the castle tower in the picture.
- Circle the solid shapes that you use to build the tower.

Learning Target: Build and explore three-dimensional shapes.

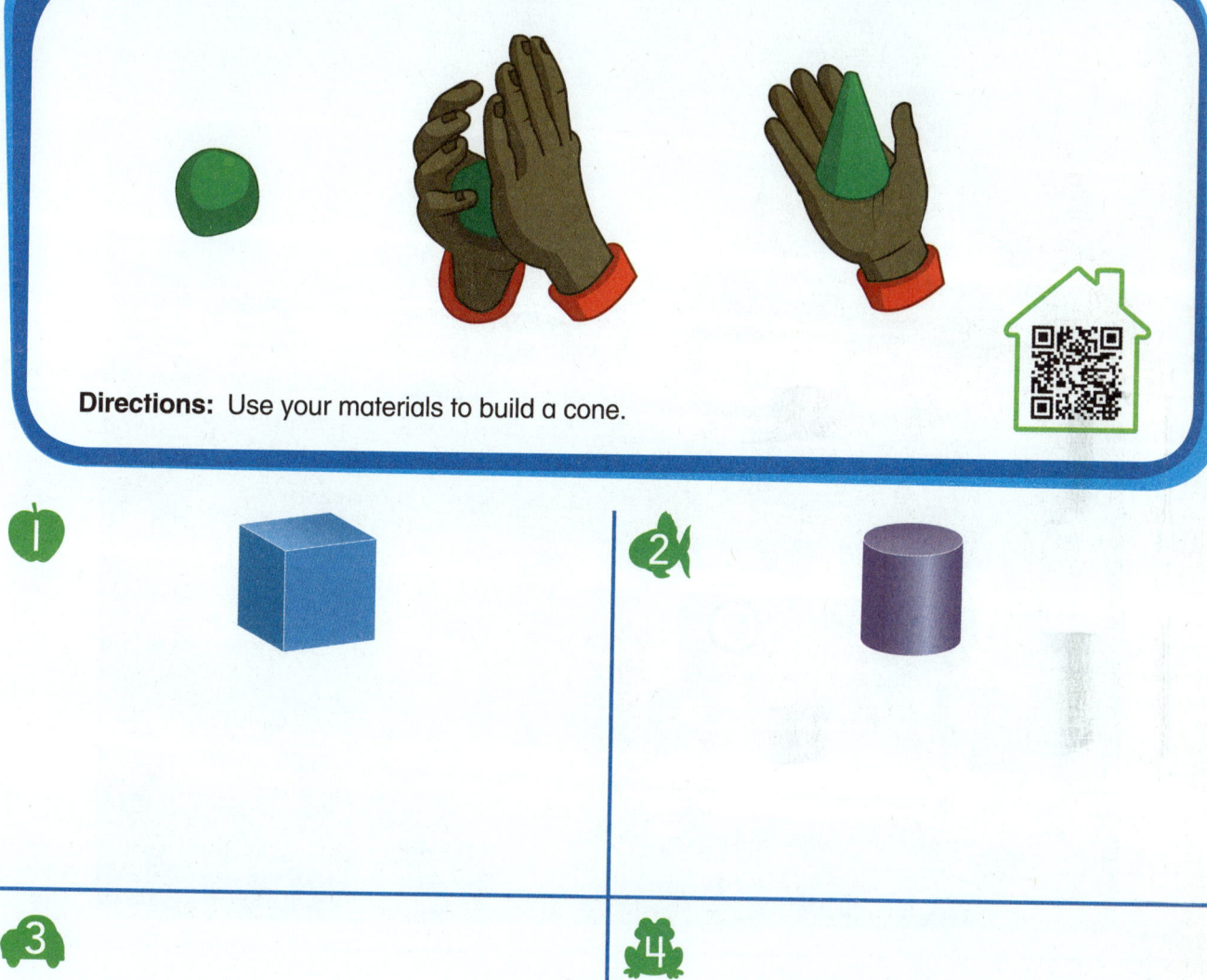

Directions: Use your materials to build a cone.

🍏 1

2 🐟

🚗 3

🐸 4

Directions: 1 and 2 Use your materials to build the solid shape shown. 3 Use your materials to build the solid shape that has a curved surface and only 1 flat surface. Circle the shape you build. 4 Use your materials to build a solid shape that has no flat surfaces. Circle the shape you build.

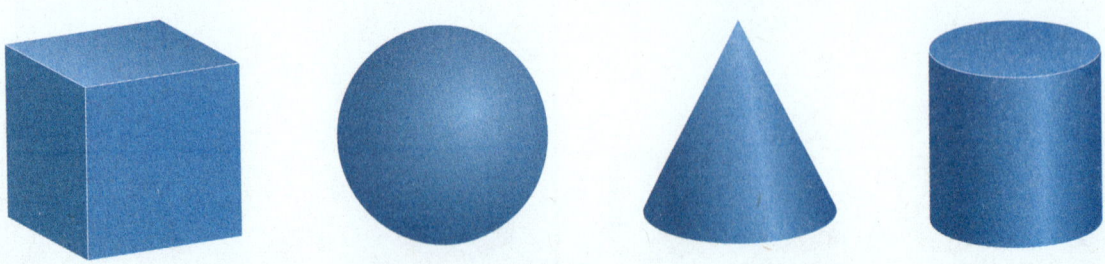

Directions: ⑤ Use your materials to build the totem pole in the picture. Circle the solid shapes that you use to make the totem pole.

Name _____

Learning Target: Describe positions of solid shapes based on other objects.

 **Explore and Grow**

Directions: Place a counter *beside* the bench. Place a counter *in front of* the tree. Place a counter *next to* the stairs. Place a counter *below* the baby swing.

Chapter 12 | **Lesson 6**

six hundred twenty-seven 627

Think and Grow

Directions:

- Circle the object that looks like a cylinder that is *next to* the table. Draw a line through the object that looks like a cone that is *below* the shelf. Draw a rectangle around the object that looks like a sphere that is *above* the table.

- Circle the object that looks like a cube that is *behind* the shovel. Draw a line through the object that looks like a cylinder that is *beside* the tree. Draw a rectangle around the object that looks like a sphere that is *in front of* the tree.

✓ Apply and Grow: Practice

Directions:  Circle the object that looks like a cylinder that is *behind* a paper cup. Draw a line through the object that looks like a sphere that is *above* the napkin dispenser. Draw a rectangle around the object that looks like a cone that is *below* a glass cup. Circle the object that looks like a cone that is *beside* the log. Draw a line through the object that looks like a sphere that is *above* the log. Draw a rectangle around the object that looks like a cone that is *in front of* the log.

 # Think and Grow: Modeling Real Life

Directions: Use the City Scene Cards to place the objects on the picture.

- Place a dog *in front of* the boy crossing the street.
- Place a tree *beside* the building that looks like a cube.
- Place an object that looks like a sphere *above* the buildings. Place that object *behind* a cloud.
- Place an object that looks like a cone *below* the traffic light.
- Place a streetlight *next to* the girl on the sidewalk.

Learning Target: Describe positions of solid shapes based on other objects.

Directions: Circle the object that looks like a cube that is *next to* the table. Draw a line through the object that looks like a cylinder that is *below* the chair. Draw a rectangle around the object that looks like a cone that is *behind* the chair.

1

Directions: **1** Circle the object that looks like a sphere that is *beside* the pool. Draw a line through the object that looks like a cone that is *next to* the ball. Draw a rectangle around the object that looks like a cylinder that is *behind* the block.

Directions: Circle the object that looks like a cone that is *above* the stuffed animal. Draw a line through the object that looks like a cylinder that is *in front of* the stuffed animal. Draw a rectangle around the object that looks like a cube that is *below* the stuffed animal.

 Use the Construction Scene Cards to place the objects on the picture. Place a building *below* the object that is shaped like a cube. Place a tree *beside* that building. Place a blimp *above* the traffic cone. Place a truck *in front of* the traffic cone.

Performance Task 12

Directions: You pick up trash in the park. Draw lines to match each item with its correct recycling bin.

- The object that rolls but does not stack that is *in front of* the lamppost goes in the yellow bin.

- The object *below* the bench that does not roll goes in the blue bin.

- The object that has 1 flat surface that is *behind* an object that looks like a cylinder goes in the green bin.

- The object that stacks, slides, and rolls that is *above* an object that looks like a cube goes in the orange bin.

- The object *in front of* the tree that rolls and has 2 flat surfaces goes in the green bin.

- The object *next to* the tree that stacks and slides and has only flat surfaces goes in the green bin.

- The object that has a curved surface that does not stack that is *beside* the tree goes in the blue bin.

- The object that slides and rolls that is *next to* an object that has 6 flat surfaces goes in the blue bin.

Solid Shapes: Spin and Cover

Directions: Take turns using the spinner to find which type of three-dimensional shape to cover. Use a counter to cover an object on the page. Repeat this process until you have covered all of the objects.

Name _____

12.1 Two- and Three-Dimensional Shapes

12.2 Describe Three-Dimensional Shapes

Directions: 1 and 2 Circle any three-dimensional shapes. Draw rectangles around any two-dimensional shapes. Tell why your answers are correct. 3 Look at the solid shape on the left that rolls. Circle the other solid shapes that roll. 4 Circle the solid shapes that stack and slide.

12.3 Cubes and Spheres

 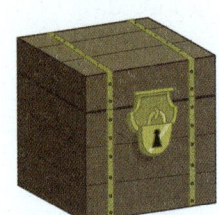  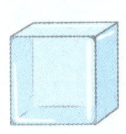

12.4 Cones and Cylinders

  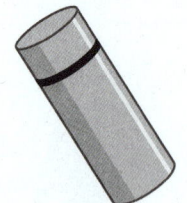 

Directions: 5 Circle the cube. Draw a rectangle around the sphere. Tell why your answers are correct. 6 Circle any object that looks like a cube. Draw a rectangle around any object that looks like a sphere. Tell why your answers are correct. 7 and 8 Circle any object that looks like a cone. Draw a rectangle around any object that looks like a cylinder. Tell why your answers are correct.

12.5 **Build Three-Dimensional Shapes**

Directions: Use your materials to build the solid shape shown. Use your materials to build a shape that has a curved surface and 2 flat surfaces. Circle the shape you build. Use your materials to build the elf in the picture. Circle the solid shapes that you use to make the elf.

 **Positions of Solid Shapes**

Directions: Circle the object that looks like a cylinder that is *below* the hat. Draw a line through the object that looks like a cone that is *beside* the cooler. Draw a rectangle around the object that looks like a cylinder that is *in front of* the hat. Circle the object that looks like a sphere that is *above* the cone. Draw a line through the object that looks like a cylinder that is *next to* the cone. Draw a rectangle around the object that looks like a sphere that is *behind* the cone.

638   six hundred thirty-eight

13

Measure and Compare Objects

Chapter Learning Target:
Understand measurement.

Chapter Success Criteria:
- ☐ I can describe height.
- ☐ I can describe weight.
- ☐ I can compare the capacities of objects.
- ☐ I can compare the heights of objects.

- Have you ever used a bucket to catch rainwater?
- Which bucket can hold the most rainwater?

13

Name ______________________

Vocabulary

Review Words
fewer
more

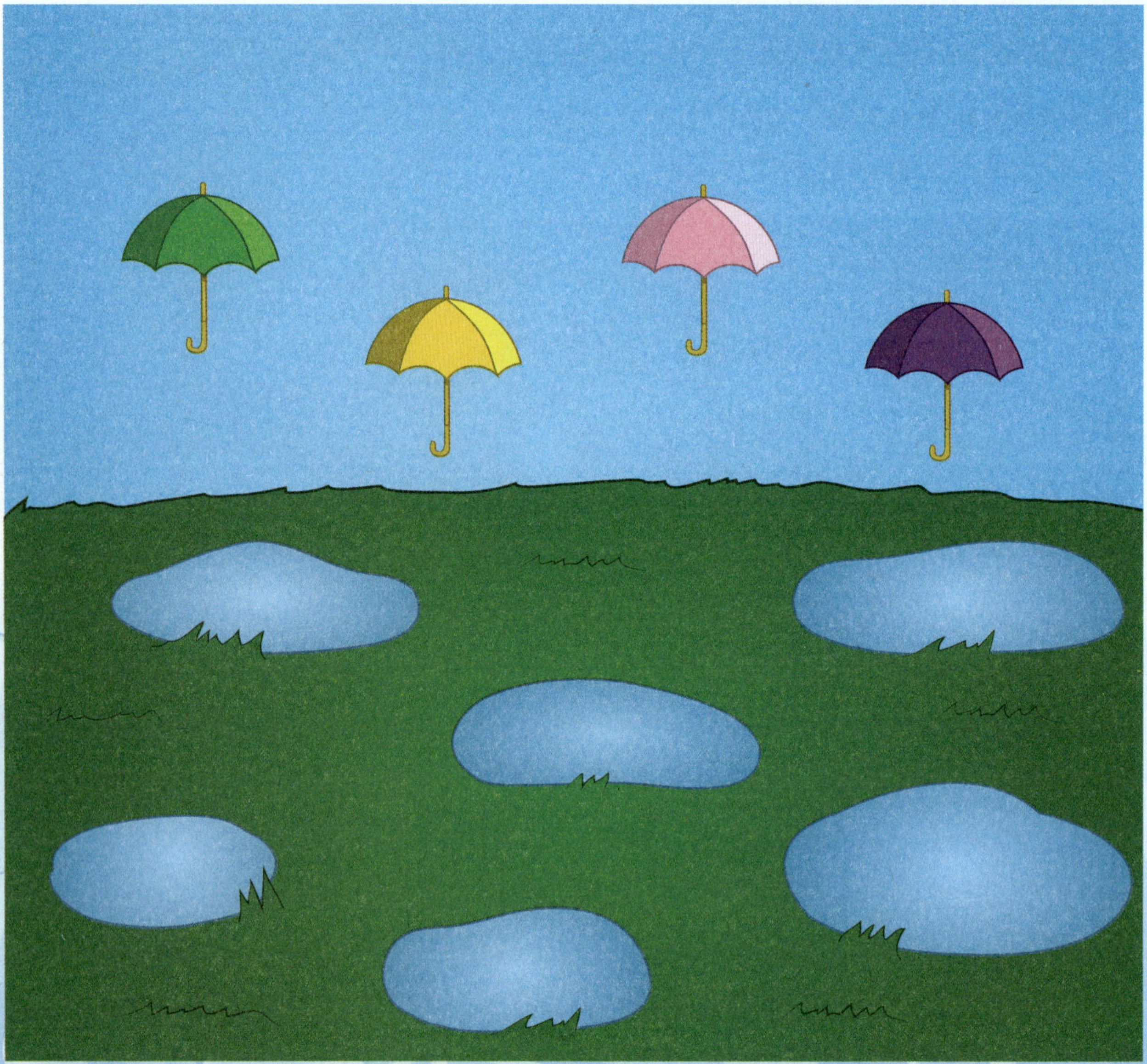

Directions: There are fewer clouds than umbrellas. There are more raindrops than puddles. Draw the clouds and the raindrops.

Chapter 13 Vocabulary Cards

balance scale

capacity

heavier

height

length

lighter

longer

measurable attribute

2 CUPS

1 CUP

Length or Height

Weight

Capacity

2 CUPS

1 CUP

shorter

taller

weight

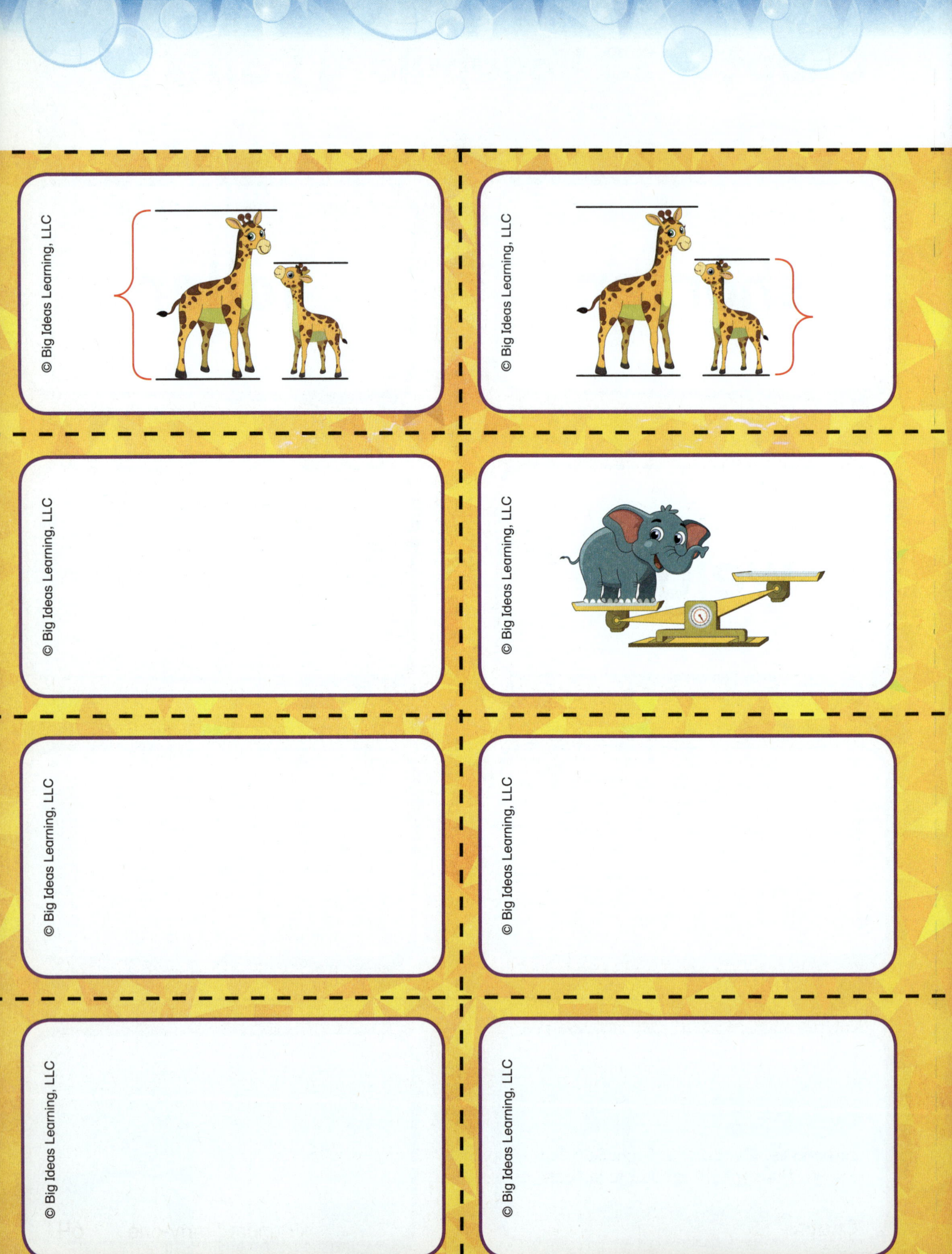

Name _____

Learning Target: Compare the heights of two objects.

 **Explore and Grow**

shorter

taller

Directions: Cut out the Height Sort Cards. Compare the objects to the children shown. Then sort the cards into the categories shown.

© Big Ideas Learning, LLC

Think and Grow

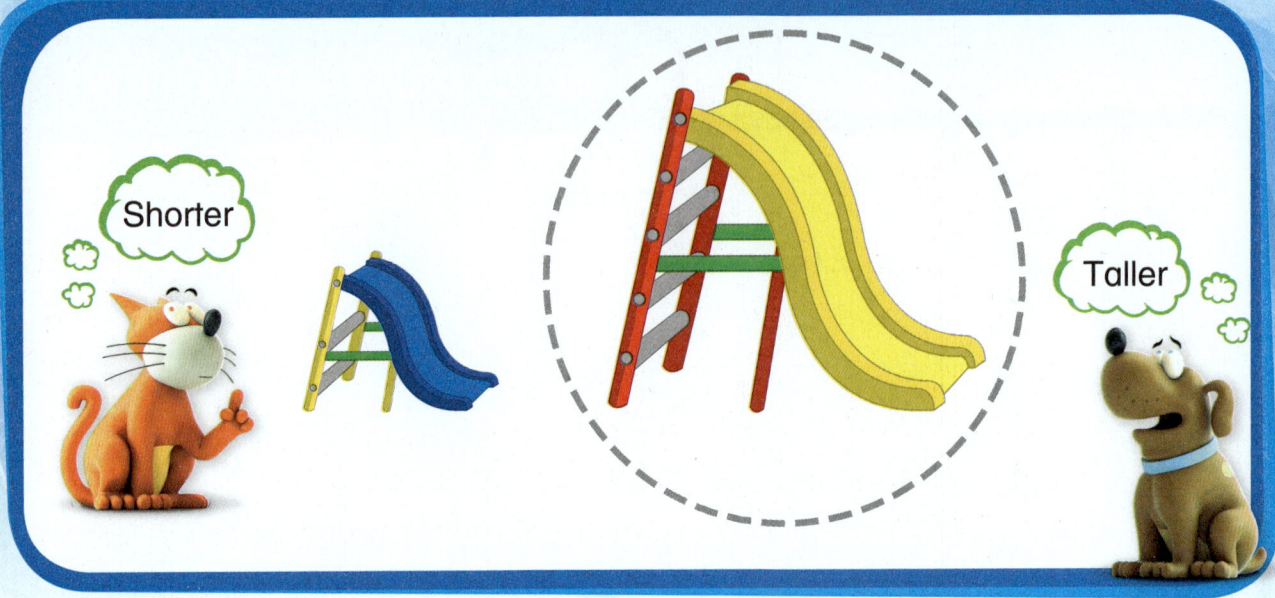

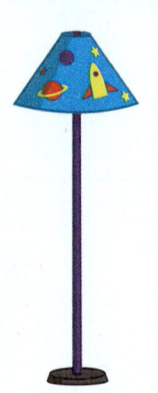

Directions: Compare the heights of the objects.
- Circle the taller slide.
- Draw a line through the shorter lamp.
- Are the mugs the same height? Circle the thumbs up for *yes* or the thumbs down for *no*.

642   six hundred forty-two

Name _____

✔ Apply and Grow: Practice

 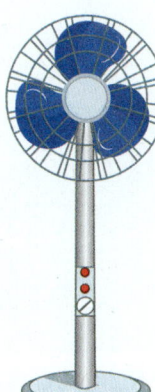

 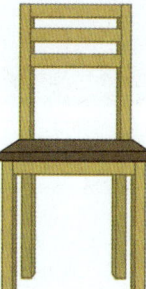

Directions: ① and ② Circle the taller object. ③ and ④ Draw a line through the shorter object. ⑤ Are the lion and the giraffe the same height? Circle the thumbs up for *yes* or the thumbs down for *no*.

Chapter 13 | Lesson 1

six hundred forty-three **643**

© Big Ideas Learning, LLC

Think and Grow: Modeling Real Life

Directions:
- Draw a building that is taller than the building shown.
- Draw a building that is shorter than the building shown.

644   six hundred forty-four

Learning Target: Compare the heights of two objects.

Directions:
• Circle the taller ladder.
• Draw a line through the shorter plant.

Directions: ❶ and ❷ Circle the taller object. ❸ and ❹ Draw a line through the shorter object.

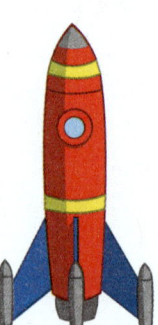 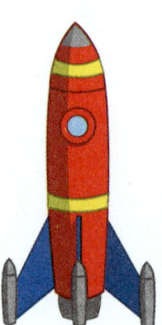

Directions: and Are the objects the same height? Circle the thumbs up for *yes* or the thumbs down for *no*. Draw a building that is the same height as the building shown.

Name _____

Learning Target: Compare the lengths of two objects.

 **Explore and Grow**

shorter

longer

Directions: Cut out the Length Sort Cards. Compare the objects to the pencil shown. Then sort the cards into the categories shown.

Directions: Compare the lengths of the objects.

- Circle the longer surfboard.
- Draw a line through the shorter watch.
- Are the shoes the same length? Circle the thumbs up for *yes* or the thumbs down for *no*.

Name _____

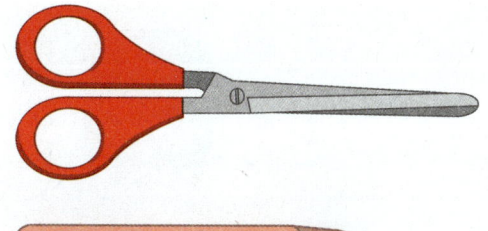

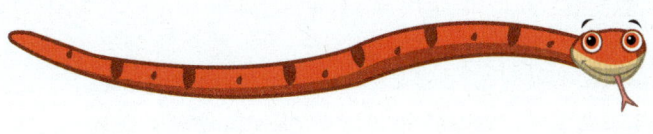

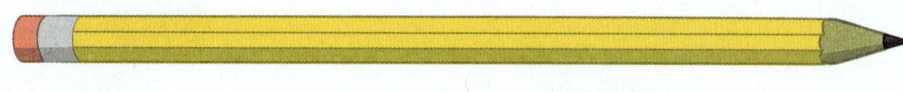

Directions: ❶ and ❷ Circle the longer object. ❸ and ❹ Draw a line through the shorter object.

Directions:
- Draw a string that holds fewer beads than the string shown. Tell how you know.
- Draw a string that holds the same number of beads as the string shown. Tell how you know.

Learning Target: Compare the lengths of two objects.

Directions:
- Circle the longer caterpillar.
- Draw a line through the shorter branch.

 **1**

2

3

 4

Directions: **1** and **2** Circle the longer object. **3** and **4** Draw a line through the shorter object.

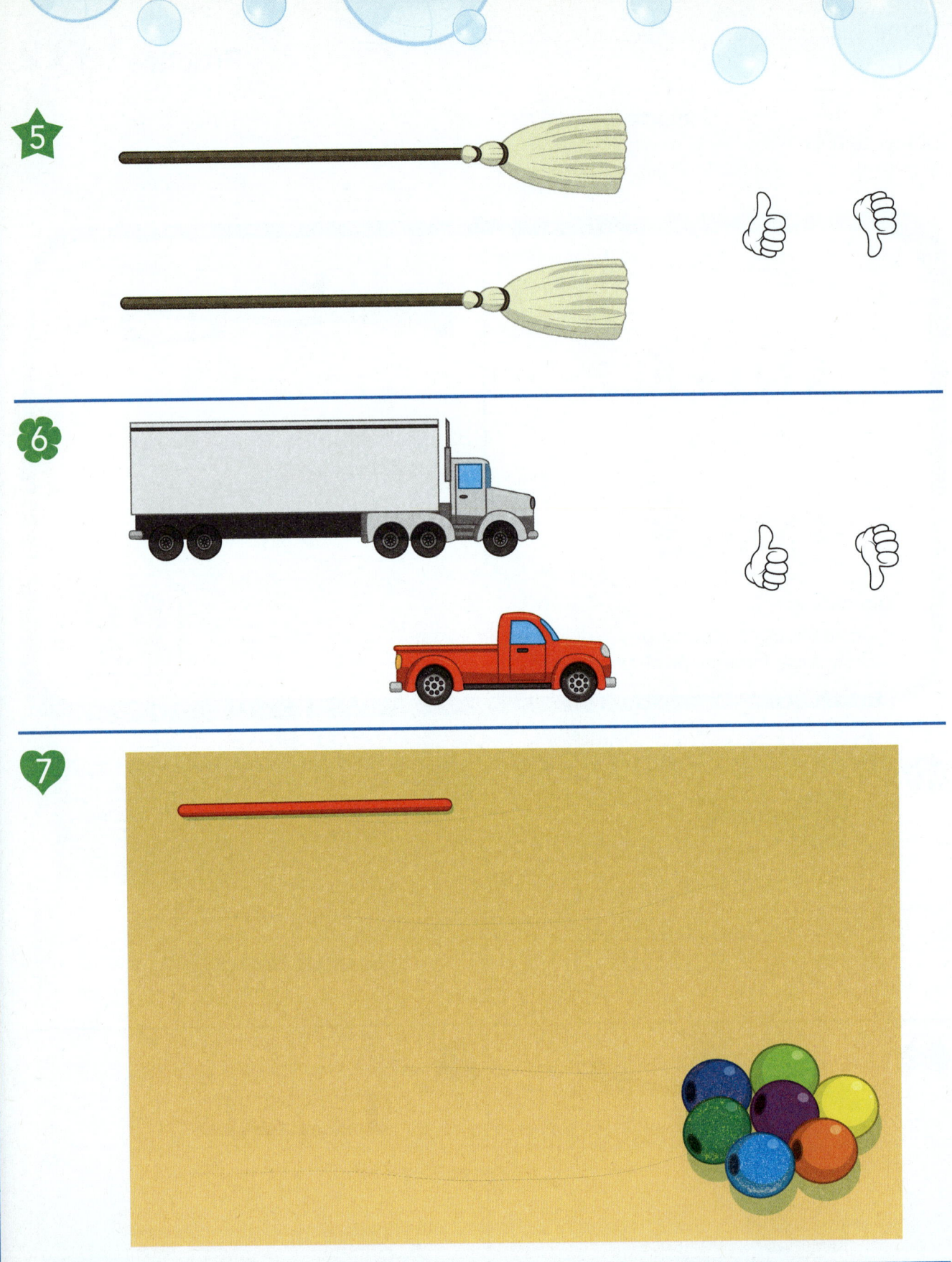

Directions: ⑤ and ⑥ Are the objects the same length? Circle the thumbs up for *yes* or the thumbs down for *no*. ⑦ Draw a string that holds more beads than the string shown. Tell how you know.

Name _____

Learning Target: Compare the lengths of two objects using numbers.

 Explore and Grow

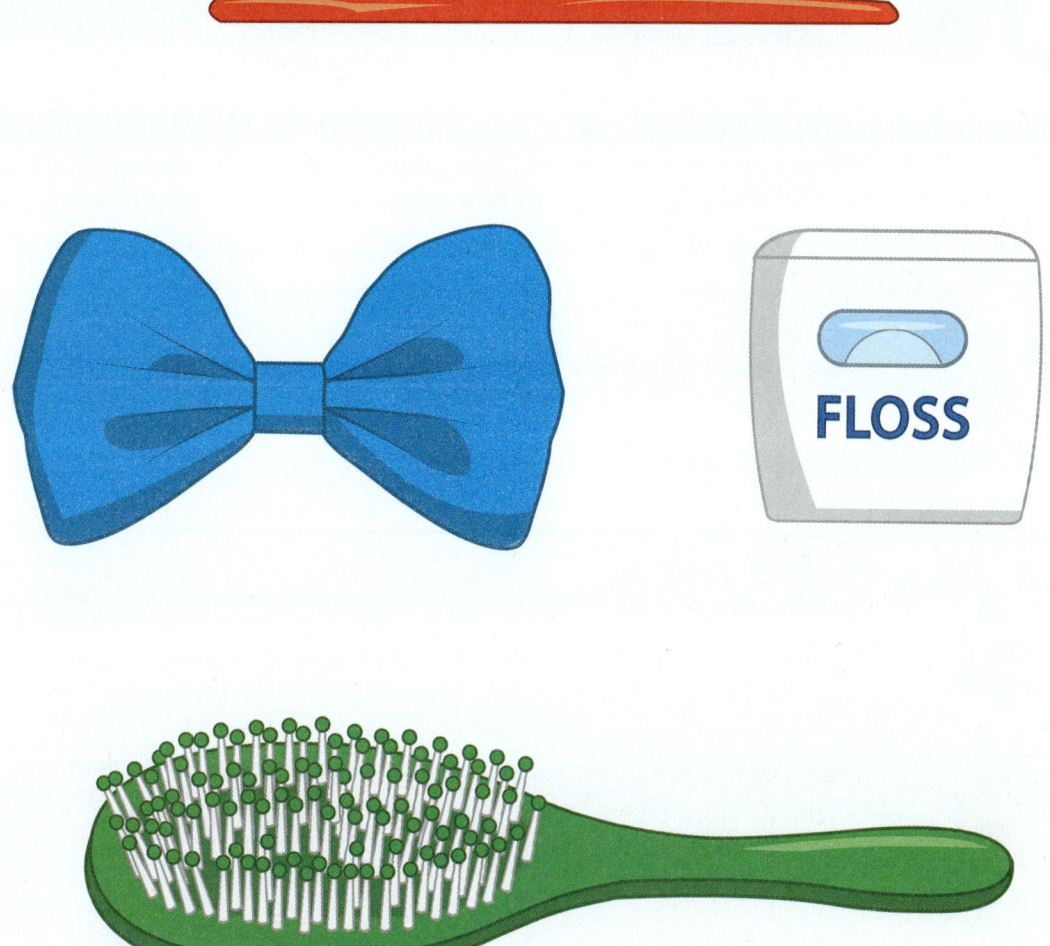

Directions: Build a linking cube train with 4 cubes. Circle the objects that are longer than the cube train.

 **Think and Grow**

2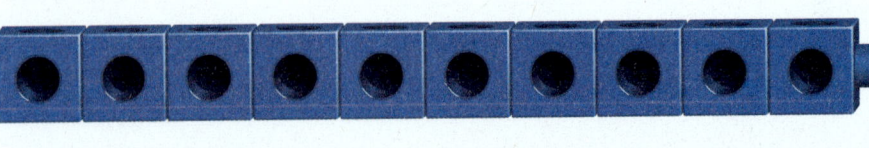

(**10**)

8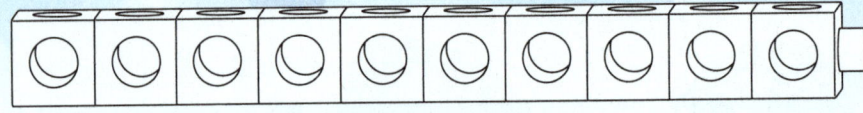

5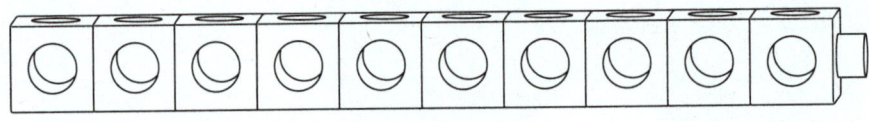

9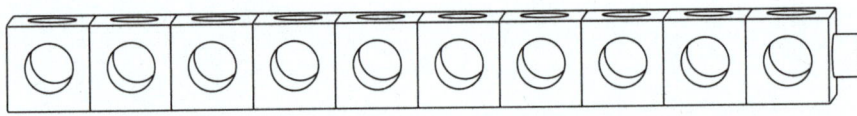

9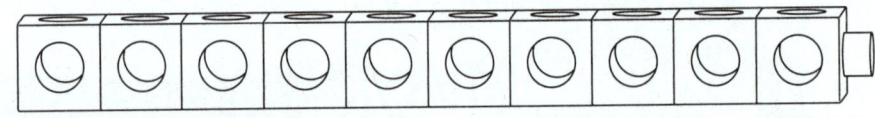

Directions: Compare the lengths of the cube trains with the given number of cubes.

• Circle the number of the train that is longer. Color to show how you know.

• Draw a line through the number of the train that is shorter. Color to show how you know.

• Are the cube trains the same length? Circle the thumbs up for *yes* or the thumbs down for *no*. Color to show how you know.

654   six hundred fifty-four

✔ Apply and Grow: Practice

10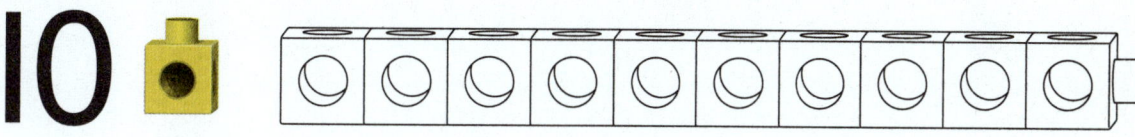

6

3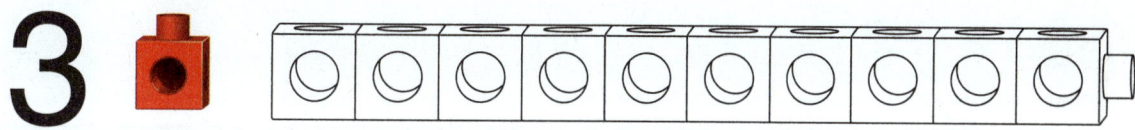

5

8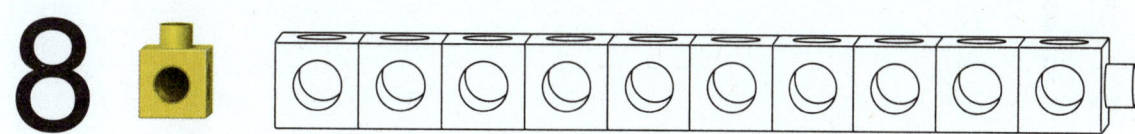

7

Directions: Compare the lengths of the cube trains with the given number of cubes. ❶ and ❷ Circle the number of the train that is longer. Color to show how you know. ❸ Draw a line through the number of the train that is shorter. Color to show how you know.

Directions: Each car on a roller coaster holds 2 people.

• Do the roller-coaster trains hold the same number of people? Circle the thumbs up for *yes* or the thumbs down for *no*. Tell how you know.

• Circle the roller-coaster train that holds more people. Tell how you know.

656   six hundred fifty-six

4

(9)

Directions: Compare the lengths of the cube trains with the given number of cubes. Circle the number of the train that is longer. Color to show how you know.

 1

3

8

2

5

1

Directions: **1** and **2** Compare the lengths of the cube trains with the given number of cubes. Circle the number of the train that is longer. Color to show how you know.

5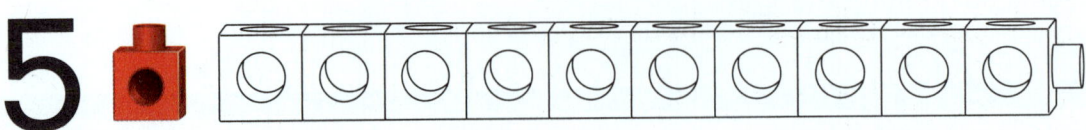

7

4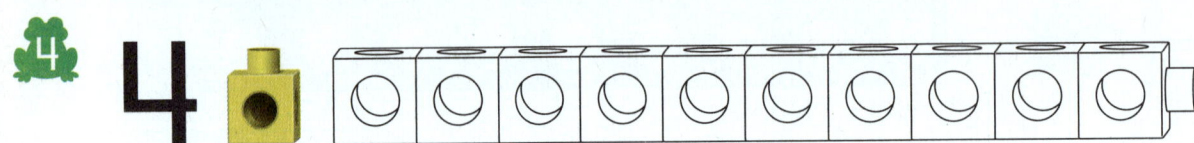

3

👍 👎

⭐ 5

Directions: 🐸3 Compare the lengths of the cube trains with the given number of cubes. Draw a line through the number of the train that is shorter. Color to show how you know. 🐸 Compare the lengths of the cube trains with the given number of cubes. Are the cube trains the same length? Circle the thumbs up for *yes* or the thumbs down for *no*. Color to show how you know. ⭐5 Each car on a roller coaster holds 2 people. Draw a line through the roller-coaster train that holds fewer people. Tell how you know.

Learning Target: Compare the weights of two objects.

Explore and Grow

lighter

heavier

Directions: Cut out the Weight Sort Cards. Compare the objects to the lion shown. Then sort the cards into the categories shown.

Think and Grow

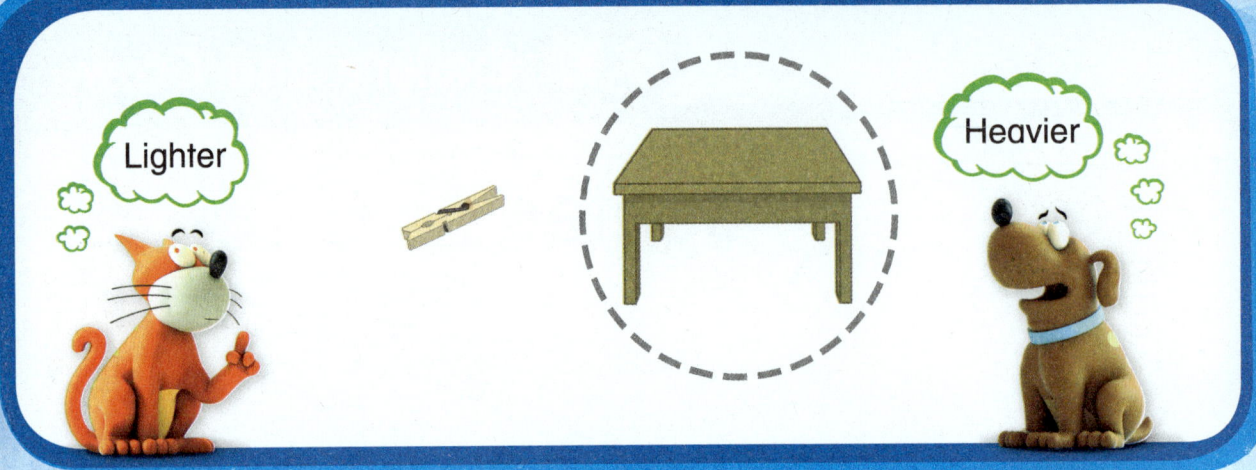

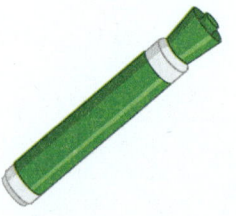   

Directions: Compare the weights of the objects.

• Circle the heavier object.

• Draw a line through the lighter object.

• Are the markers the same weight? Circle the thumbs up for *yes* or the thumbs down for *no*.

Apply and Grow: Practice

  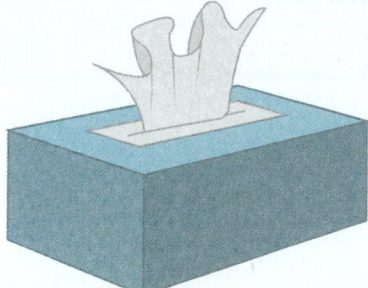  

Directions: 1 and 2 Circle the heavier object. 3 and 4 Draw a line through the lighter object. 5 Are the objects the same weight? Circle the thumbs up for *yes* or the thumbs down for *no*.

Chapter 13 | **Lesson 4**

Think and Grow: Modeling Real Life

Directions:

• Circle the object you can carry. Tell why you can carry the object.

• Circle the object you *cannot* carry. Tell why you *cannot* carry the object.

Learning Target: Compare the weights of two objects.

 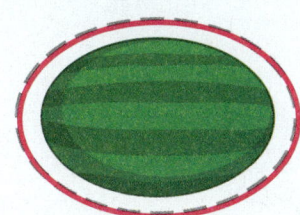  

Directions:
- Circle the heavier fruit.
- Draw a line through the lighter toy.

 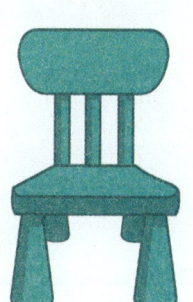

 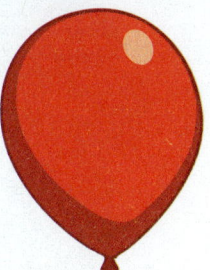

Directions: and Circle the heavier object. and Draw a line through the lighter object.

Chapter 13 | Lesson 4

5

  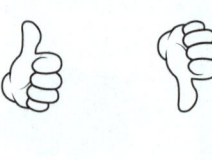

6

7

Directions: **5** and **6** Are the objects the same weight? Circle the thumbs up for *yes* or the thumbs down for *no*. **7** Circle the object you can carry. Tell why you can carry the object.

664   six hundred sixty-four

© Big Ideas Learning, LLC

Name _____

Learning Target: Compare the weights of two objects using numbers.

Explore and Grow

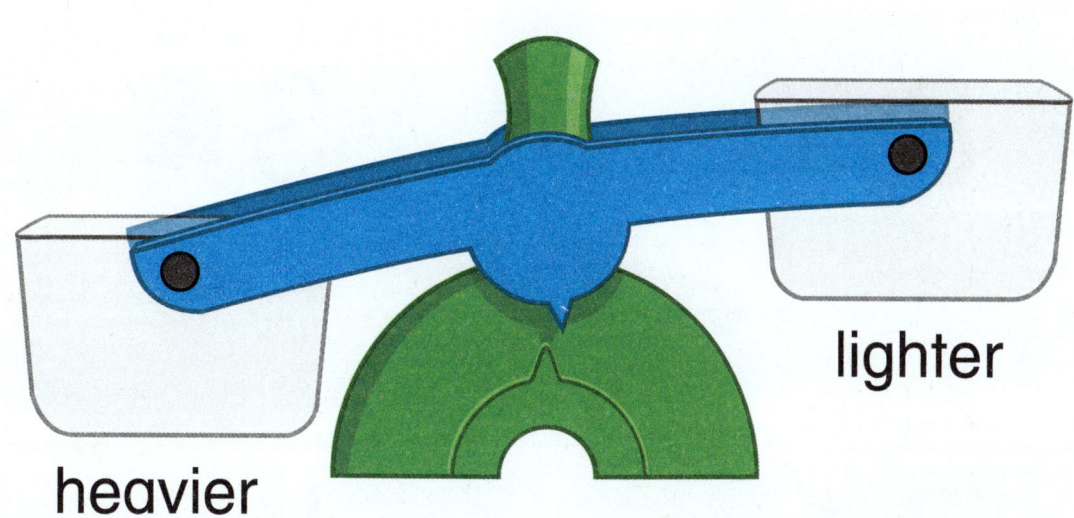

lighter

heavier

Directions: Hold some counting bears in one hand and a different amount of counting bears in your other hand. Place the groups of bears on the correct buckets on the scale.

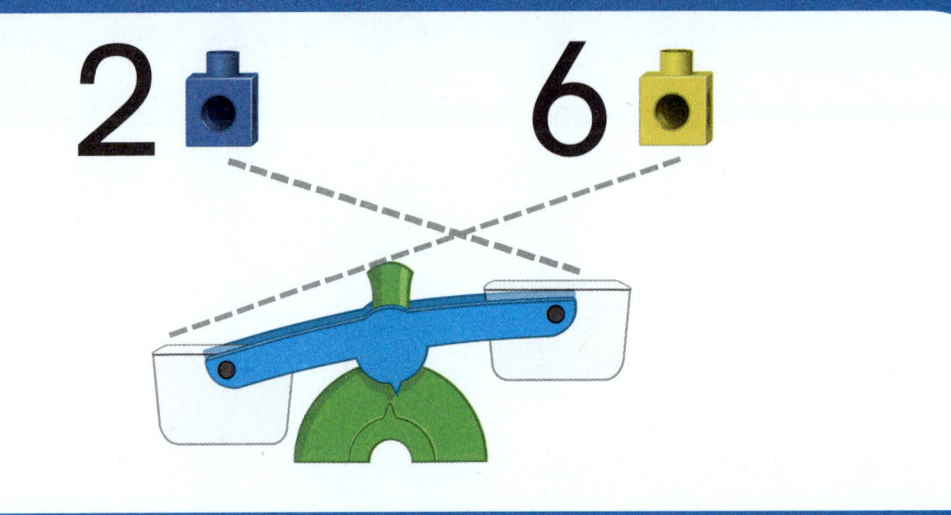

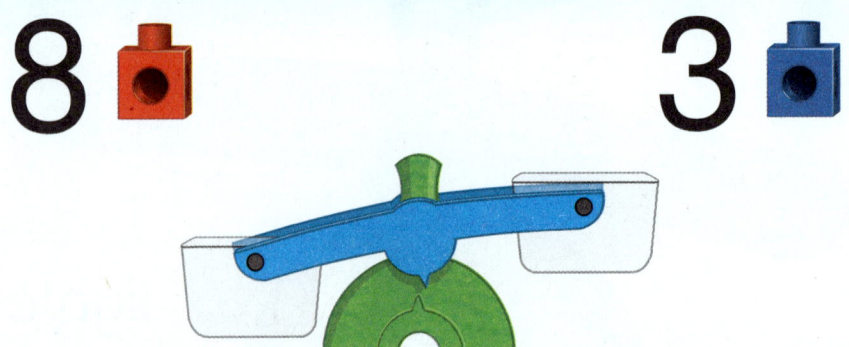

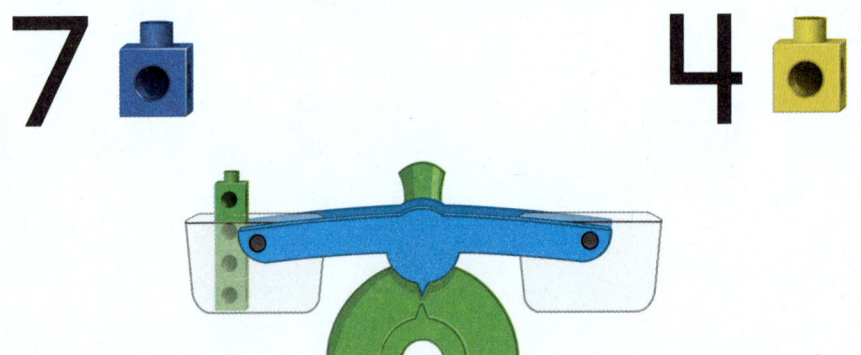

Directions:

- Compare the weights of the groups of linking cubes. Match each group of linking cubes with the correct side of the balance.
- Compare the weights of the groups of linking cubes. Match each group of linking cubes with the correct side of the balance.
- Circle the number of linking cubes that makes the balance scale even.

✓ Apply and Grow: Practice

 **7**  **4**

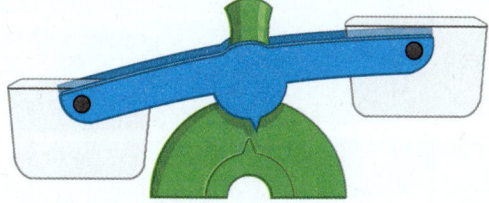

 3 **6**

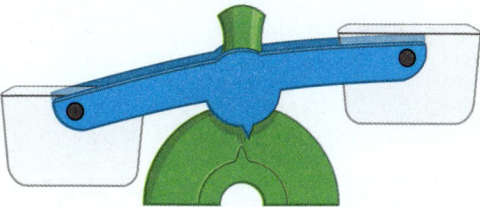

 10  **3** ☐ **+ 1** ☐

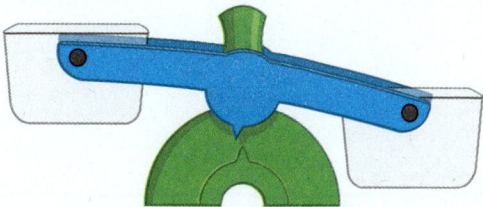

Directions: ❶–❸ Compare the weights of the groups of linking cubes. Match each group of linking cubes with the correct side of the balance scale.

Directions:
- Circle the basket that is lighter. Tell how you know.
- Circle the basket that is heavier. Tell how you know.

668   six hundred sixty-eight

Learning Target: Compare the weights of two objects using numbers.

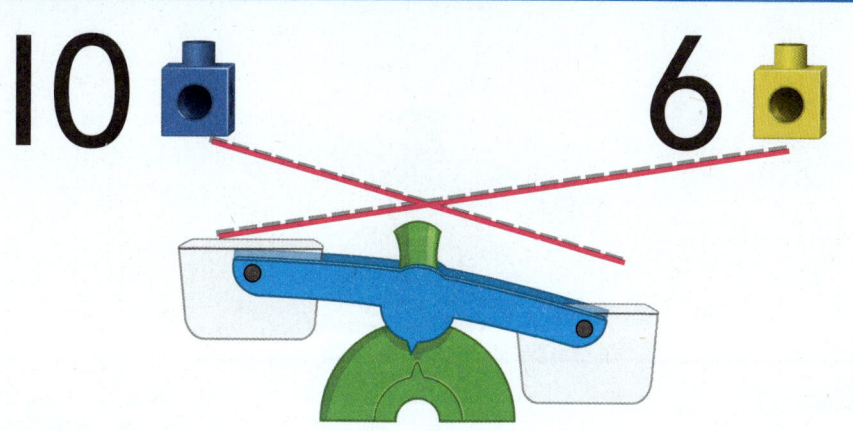

Directions: Compare the weights of the groups of linking cubes. Match each group of linking cubes with the correct side of the balance scale.

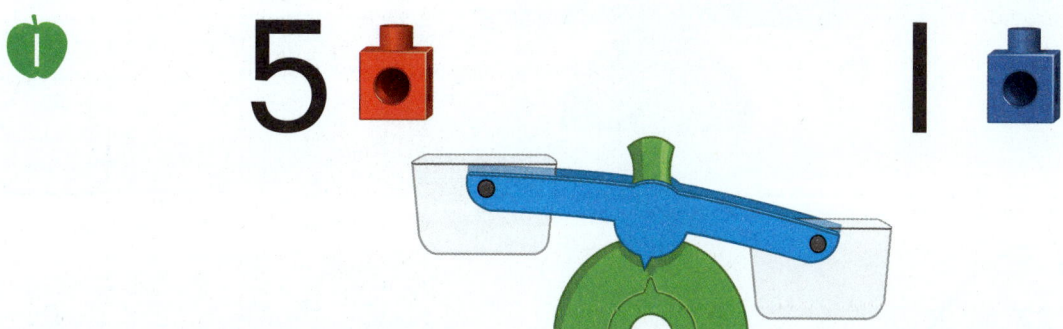

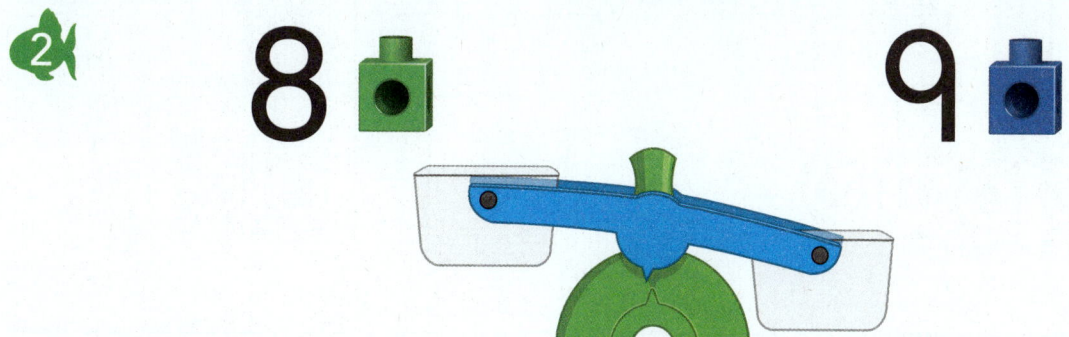

Directions: 1 and 2 Compare the weights of the groups of linking cubes. Match each group of linking cubes with the correct side of the balance scale.

Chapter 13 | Lesson 5

5    **2** + **2**

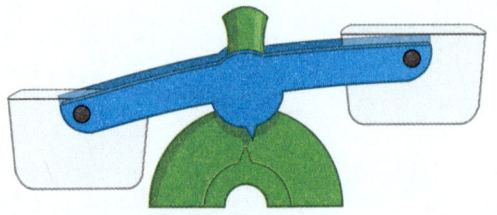

7   **3**

5

Directions: **3** Compare the weights of the groups of linking cubes. Match each group of linking cubes with the correct side of the balance scale. **4** Circle the number of linking cubes that makes the balance scale even. **5** Circle the basket that is heavier. Tell how you know.

Learning Target: Compare the capacities of two objects.

Explore and Grow

holds less

holds more

Directions: Cut out the Capacity Sort Cards. Compare the objects to the bucket shown. Then sort the cards into the categories shown.

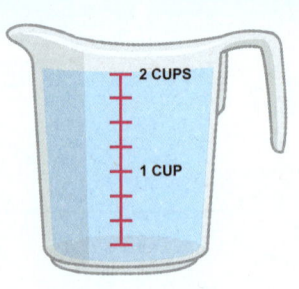

Directions: Compare the capacities of the objects.

- Circle the object that holds more.
- Draw a line through the object that holds less.
- Do the recycling bins hold the same amount? Circle the thumbs up for *yes* or the thumbs down for *no*.

Name _____

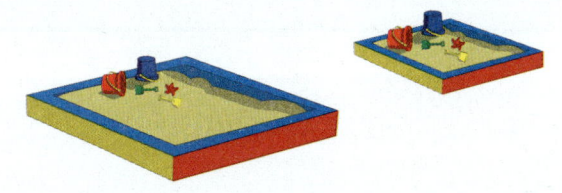

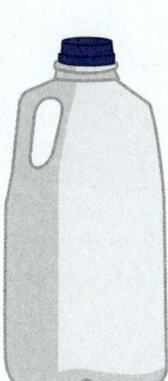   

Directions:  and Circle the object that holds more. and  Draw a line through the object that holds less. Do the milk containers hold the same amount? Circle the thumbs up for *yes* or the thumbs down for *no*.

Chapter 13 | Lesson 6

Directions:

- You are going camping. Circle the backpack that can hold all of your camping supplies. Tell how you know.
- You are going to school. Circle the bag that *cannot* hold all of your school supplies. Tell how you know.

Learning Target: Compare the capacities of two objects.

Directions:
- Circle the cup that holds more.
- Draw a line through the vase that holds less.

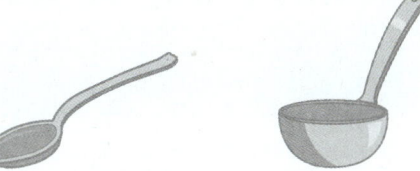

Directions: ➊ and ➋ Circle the object that holds more. ➌ and ➍ Draw a line through the object that holds less.

Chapter 13 | **Lesson 6**

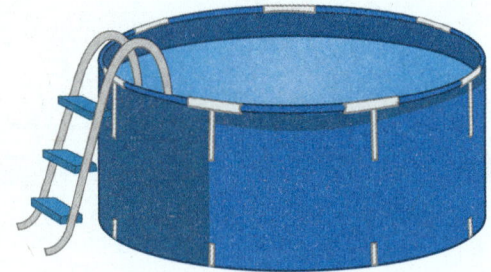   

Directions: and Do the objects hold the same amount? Circle the thumbs up for *yes* or the thumbs down for *no*. Your class is going on a field trip. Circle the vehicle that can hold your class. Tell how you know.

Name _____

Learning Target: Identify the measurable attributes of an object.

 **Explore and Grow**

length or height

weight

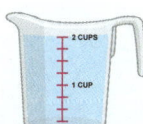

capacity

Directions: Cut out the Measurable Attribute Sort Cards. Place the objects that you can measure using length or height into the length or height box. Then place the objects that you can measure using weight into the weight box. Then place the objects that you can measure using capacity into the capacity box.

© Big Ideas Learning, LLC

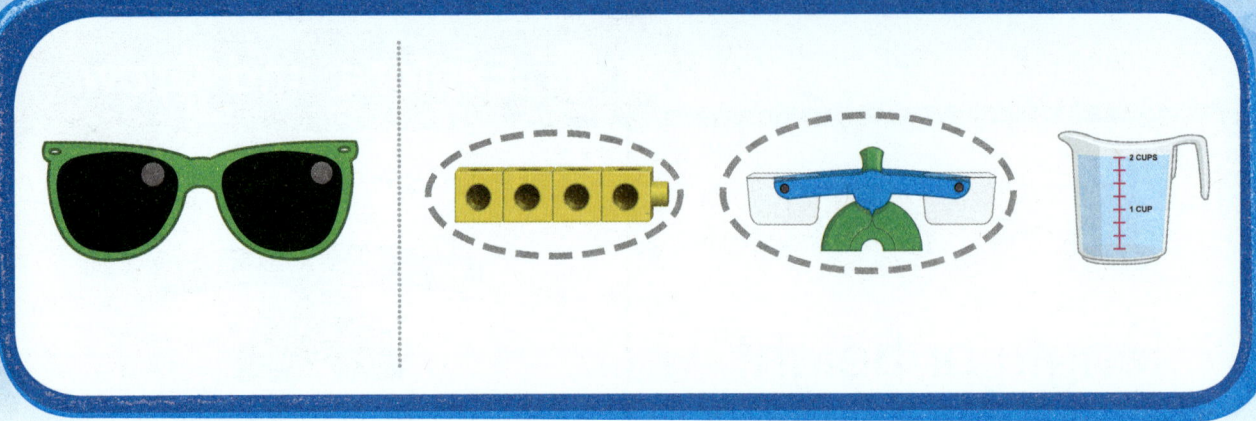

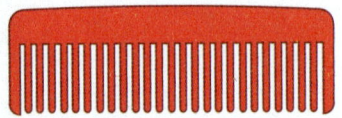   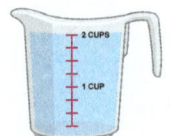

  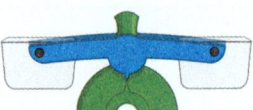 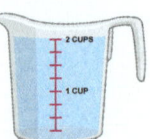

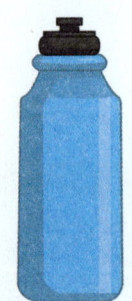  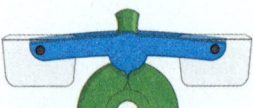 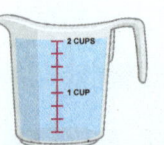

Directions: Circle the measurable attributes of the object.

✔ Apply and Grow: Practice

1  |

2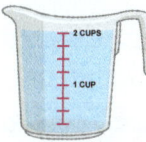

3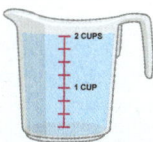

4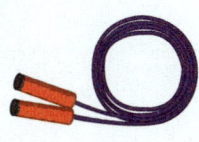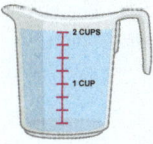

5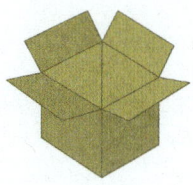

Directions: **1**–**4** Circle the measurable attributes of the object. **5** Circle the objects that have capacity as an attribute.

Chapter 13 | Lesson 7

Think and Grow: Modeling Real Life

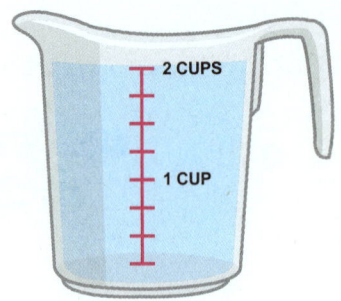

Directions:
- Draw an object that has capacity as an attribute.
- Draw an object that does *not* have capacity as an attribute.

680   six hundred eighty

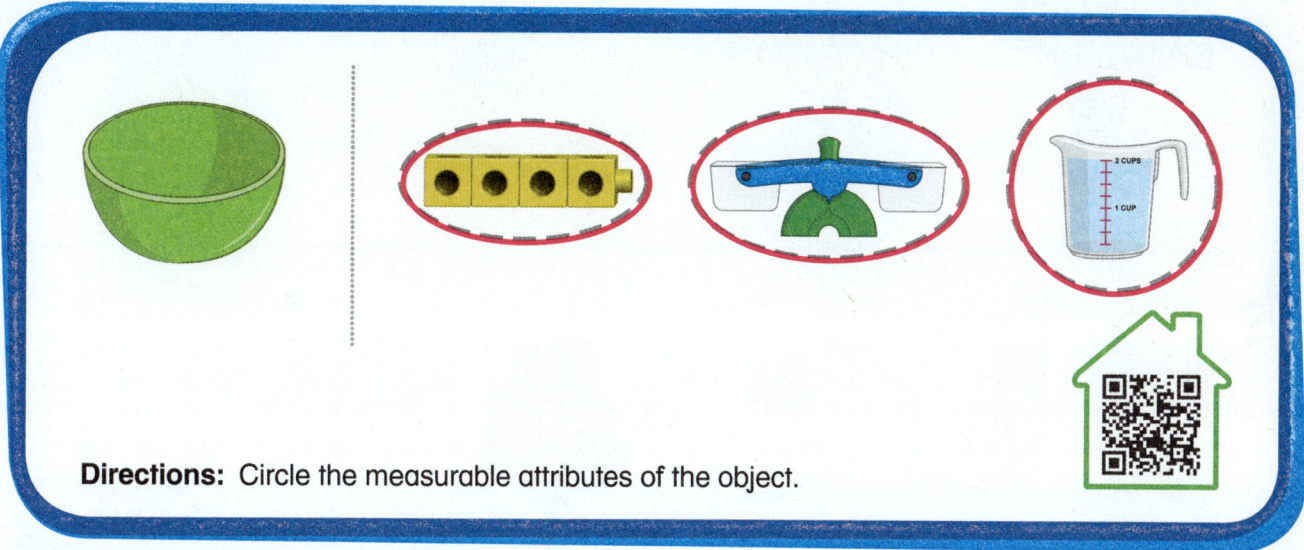

Directions: Circle the measurable attributes of the object.

1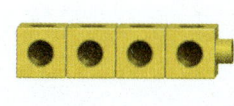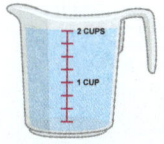

2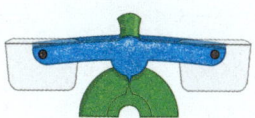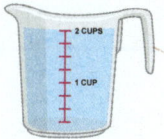

3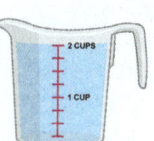

Directions: **1**–**3** Circle the measurable attributes of the object.

 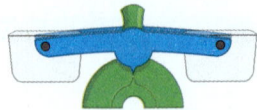 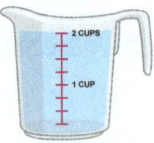

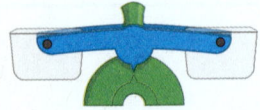

Directions: Circle the measurable attributes of the broccoli. 5 Circle the objects that have length as an attribute. 6 Draw an object that has length as an attribute. 7 Draw an object that has weight as an attribute.

Performance Task 13

Monday   Tuesday

Wednesday

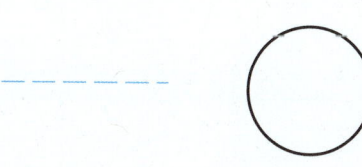 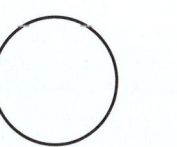 

Directions: You use one bucket to collect rainwater on Monday and a different bucket to collect rainwater on Tuesday. On Monday, you collect 1 less than 7 fluid ounces of rainwater. On Tuesday, you collect 1 more than 3 fluid ounces of rainwater. Circle the number on each bucket that shows the amount of rainwater you collect. Then circle the day that you collect more rainwater. Draw a bucket for Wednesday that is taller and holds more water than Monday's bucket. The amount of rainwater you collect on Wednesday is the same as the amount you collect in all on Monday and Tuesday. Write an addition sentence to tell how much rainwater you collect on Wednesday.

Measurement Boss

Player 1	Player 2

Directions: Each player flips a Measurement Boss Card and places it on the page. Compare the objects based on the attribute of the card. The player with the object that is longer, taller, heavier, or holds more takes both cards. Repeat until all cards have been used.

13.1 Compare Heights

13.2 Compare Lengths

  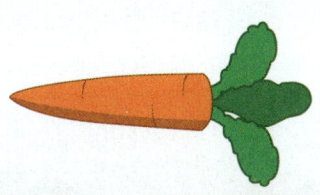

 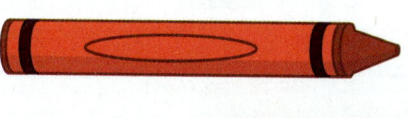

Directions: ① and ② Circle the taller object. ③ Draw a line through the shorter object. ④ Are the crayons the same length? Circle the thumbs up for *yes* or the thumbs down for *no*. ⑤ Draw a string that holds the same number of beads as the string shown. Tell how you know.

 7

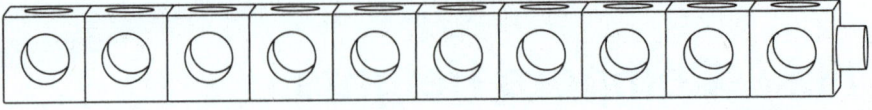

5

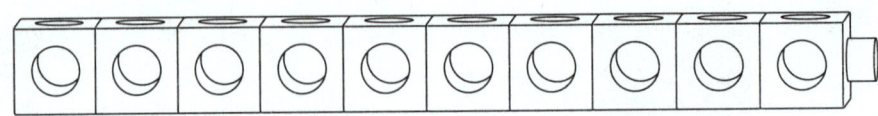

 10

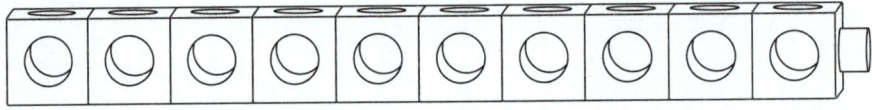

9

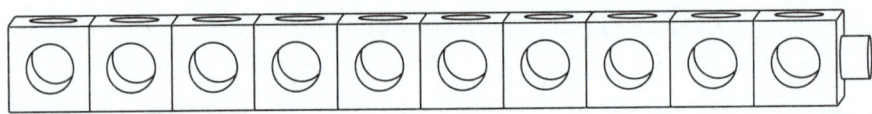

13.4 Compare Weights

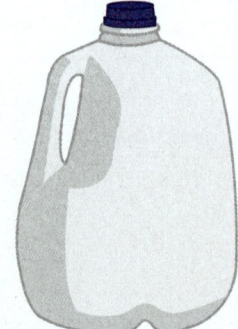

Directions: 6 and 7 Compare the lengths of the cube trains that have the given number of cubes. Circle the number of the train that is longer. Color to show how you know. 8 Draw a line through the lighter object. 9 Are the footballs the same weight? Circle the thumbs up for *yes* or the thumbs down for *no*.

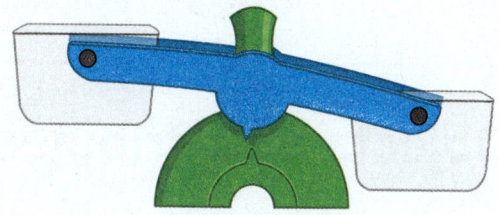

13.6 Compare Capacities

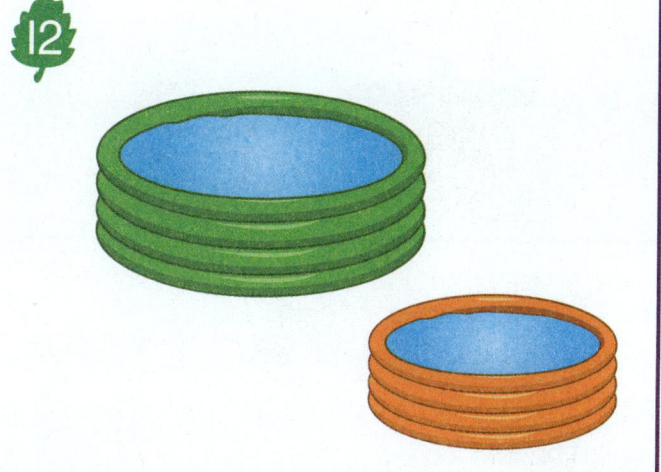  

Directions: Compare the weights of the groups of linking cubes. Match each group of linking cubes with the correct side of the balance scale. Circle the bag that is lighter. Tell how you know. 12 and 13 Draw a line through the object that holds less.

13.7 Describe Objects by Attributes

   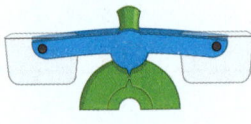 

  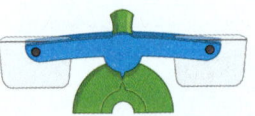 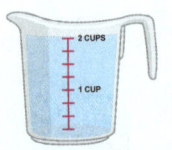

Directions: Do the water bottles hold the same amount? Circle the thumbs up for *yes* or the thumbs down for *no.* 15–17 Circle the measurable attributes of the object. 18 Circle the objects that have capacity as an attribute.

1 ◯   ◯   ◯   ◯

2 ◯ ◯ ◯ ◯

3 ◯   ◯   ◯   ◯

Directions: Shade the circle next to the answer. **1** Which group has a yellow pencil that is longer than the red pencil? **2** Which five frame shows how many sharks are in the picture? **3** Which group has all rectangles?

4

- $8 - 3 = 5$
- $4 - 3 = 1$
- $5 - 3 = 2$
- $6 - 3 = 3$

5

◯   ◯   ◯   ◯

6

◯   ◯   ◯   ◯

Directions: Shade the circle next to the answer. **4** Which subtraction sentence tells how many geese are left? **5** Which shape is *not* a solid shape? **6** Which solid shape does *not* stack or slide?

690   six hundred ninety

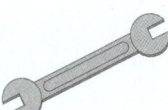

8

9

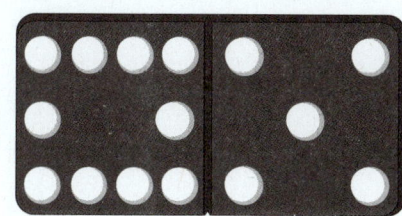
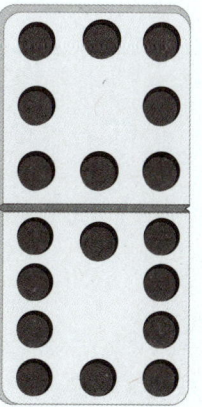

○ _____

● _____

Directions: 7 Circle the objects that have capacity as an attribute. 8 Circle the object that looks like a cylinder that is *above* the ball. 9 Find the number of dots on each domino. Write each number. Draw a line through the number that is less than the other number.

10

_____ sides   _____ vertices

11

12

Directions: **10** Trace the shapes that are hexagons. Write the number of sides and the number of vertices of a hexagon. **11** You have 10 apples. Classify the apples into 2 categories. Circle the groups. Then complete the number bond to match your picture. **12** Draw a larger triangle that can be formed by the 2 triangles shown.

Glossary

 A

above [arriba, encima]

add [sumar]

$$2 + 4 = 6$$

addition sentence
[enunciado suma]

$$2 + 3 = 5$$

 B

balance scale [balanza]

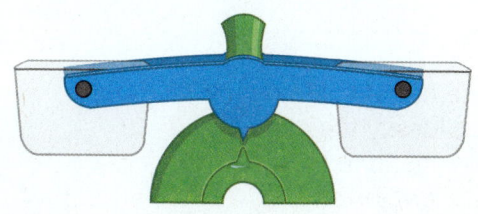

behind [detrás]

below [debajo]

beside [al lado]

capacity [capacidad]

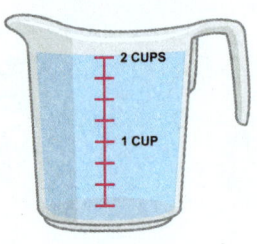

category [categoría]

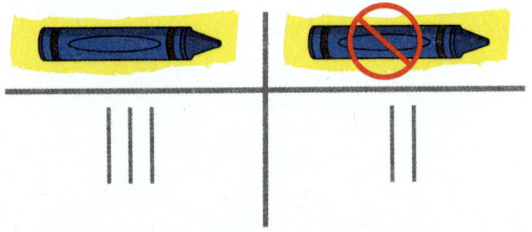

chart [gráfico]

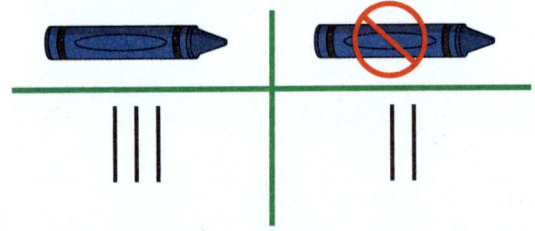

circle [círculo]

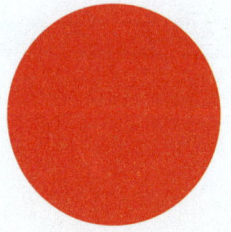

classify [clasificar]

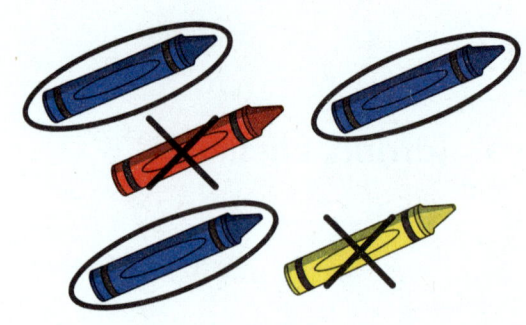

column [columna]

1	2	3	4	5	6	7	8	9	10
11	12	13	14	15	16	17	18	19	20
21	22	23	24	25	26	27	28	29	30
31	32	33	34	35	36	37	38	39	40
41	42	43	44	45	46	47	48	49	50
51	52	53	54	55	56	57	58	59	60
61	62	63	64	65	66	67	68	69	70
71	72	73	74	75	76	77	78	79	80
81	82	83	84	85	86	87	88	89	90
91	92	93	94	95	96	97	98	99	100

compare [comparar]

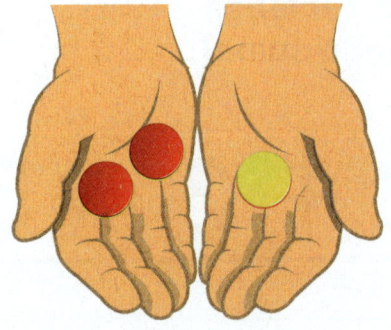

cone [cono]

count [contar]

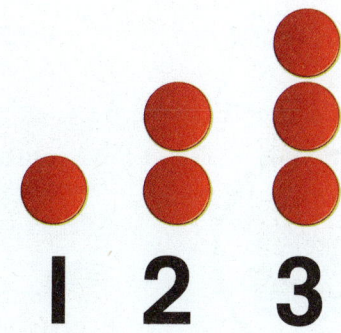

1 2 3

cube [cubo]

curve [curva]

curved surface [superficie curva]

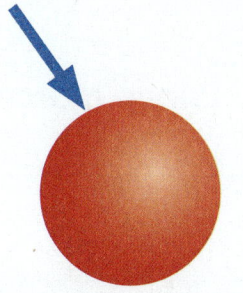

cylinder [cilindro]

D

decade number
[número de década]

1	2	3	4	5	6	7	8	9	10
11	12	13	14	15	16	17	18	19	20
21	22	23	24	25	26	27	28	29	30
31	32	33	34	35	36	37	38	39	40
41	42	43	44	45	46	47	48	49	50
51	52	53	54	55	56	57	58	59	60
61	62	63	64	65	66	67	68	69	70
71	72	73	74	75	76	77	78	79	80
81	82	83	84	85	86	87	88	89	90
91	92	93	94	95	96	97	98	99	100

E

eight [ocho]

8

A3

eighteen   [dieciocho]

18

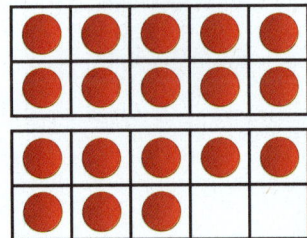

eleven   [once]

11

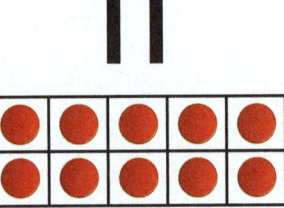

equal   [igual]

3
3

equal sign   [signo igual]

$$3 + 4 = 7$$

fewer   [menos]

fifteen   [quince]

15

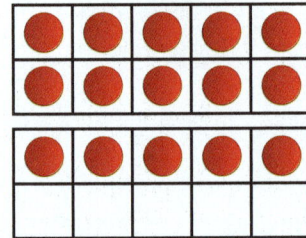

five   [cinco]

5

five frame   [cinco marco]

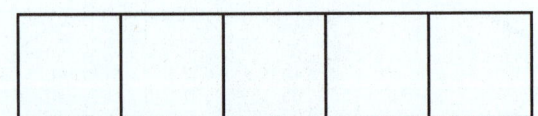

A4

flat surface [superficie plana]

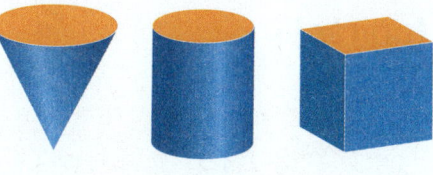

four [cuatro]

fourteen [catorce]

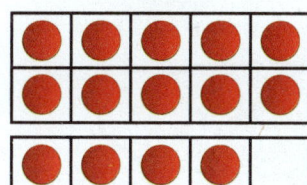

G

greater than [mas grande que]

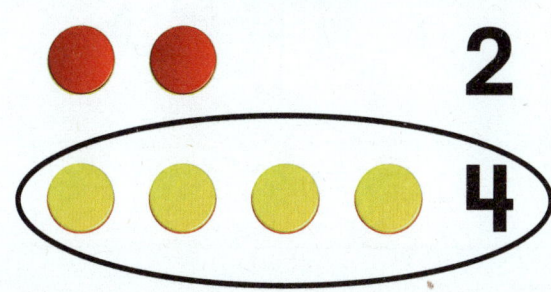

H

heavier [más pesado]

height [altura]

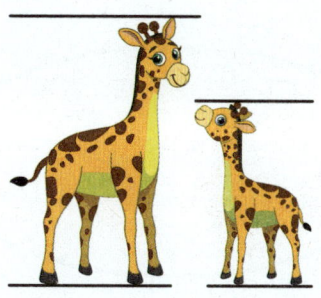

hexagon [hexágono]

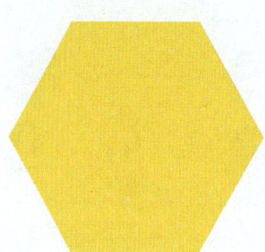

hundred chart [cientos de cartas]

1	2	3	4	5	6	7	8	9	10
11	12	13	14	15	16	17	18	19	20
21	22	23	24	25	26	27	28	29	30
31	32	33	34	35	36	37	38	39	40
41	42	43	44	45	46	47	48	49	50
51	52	53	54	55	56	57	58	59	60
61	62	63	64	65	66	67	68	69	70
71	72	73	74	75	76	77	78	79	80
81	82	83	84	85	86	87	88	89	90
91	92	93	94	95	96	97	98	99	100

I

in all [en todo]

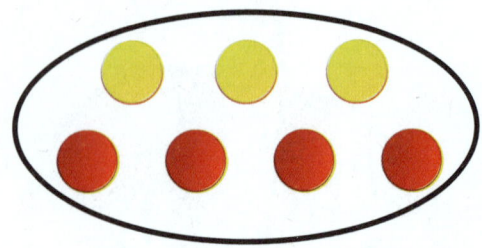

in front of [delante de]

J

join [unirse]

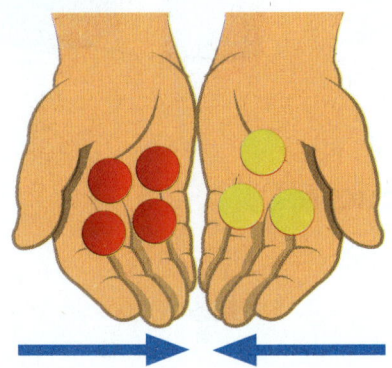

L

left [izquierda]

length [longitud]

A6

less than   [menos que]

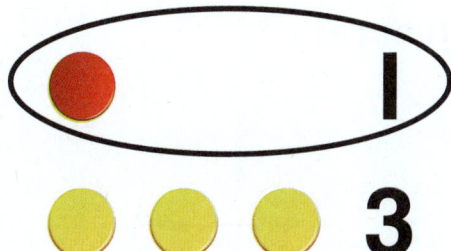

lighter   [más liviano]

longer   [más largo]

mark   [marca]

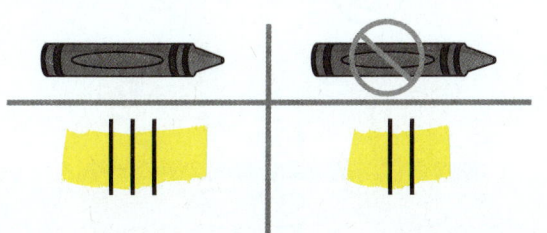

measurable attribute
[atributo mensurable]

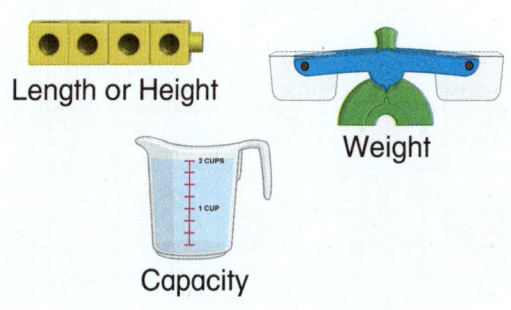

Length or Height

Weight

Capacity

minus sign   [signo menos]

$$3 - 2 = 1$$

more   [más]

next to   [al lado de]

nine [nueve]

nineteen [diecinueve]

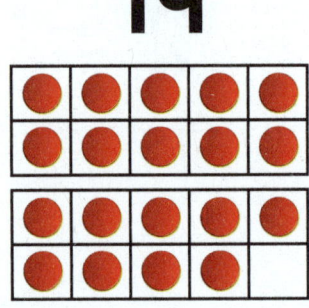

number [número]

number bond [número de bonos]

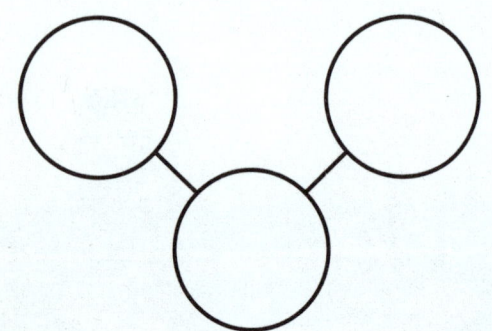

one [uno]

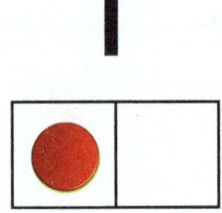

order [ordenar]

part [parte]

partner numbers
[números de socio]

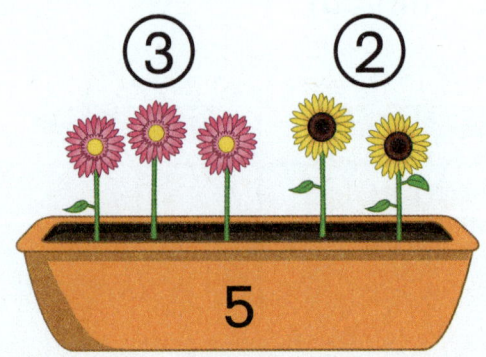

pattern   [patrón]

$$1 + 1 = 2$$
$$2 + 1 = 3$$
$$3 + 1 = 4$$

plus sign   [signo de más]

$$2 + 1 = 3$$

put together   [juntar]

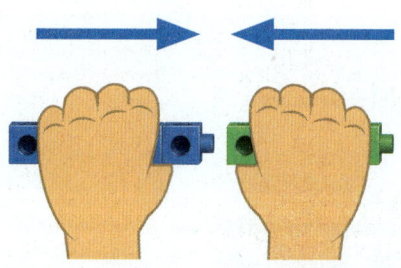

R

rectangle   [rectángulo]

roll   [rodar]

row   [fila]

1	2	3	4	5	6	7	8	9	10
11	12	13	14	15	16	17	18	19	20
21	22	23	24	25	26	27	28	29	30
31	32	33	34	35	36	37	38	39	40
41	42	43	44	45	46	47	48	49	50
51	52	53	54	55	56	57	58	59	60
61	62	63	64	65	66	67	68	69	70
71	72	73	74	75	76	77	78	79	80
81	82	83	84	85	86	87	88	89	90
91	92	93	94	95	96	97	98	99	100

S

same as   [igual que]

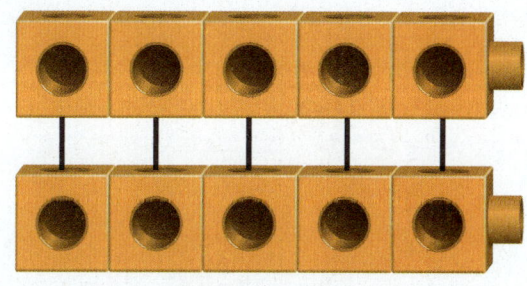

separate   [separar]

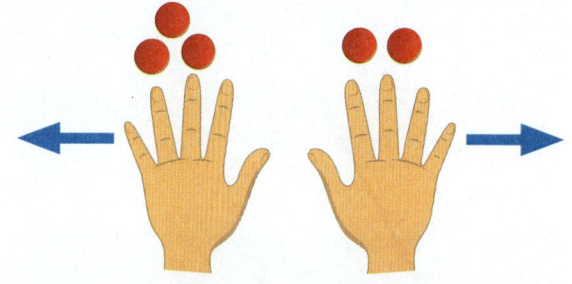

A9

seven [siete]

7

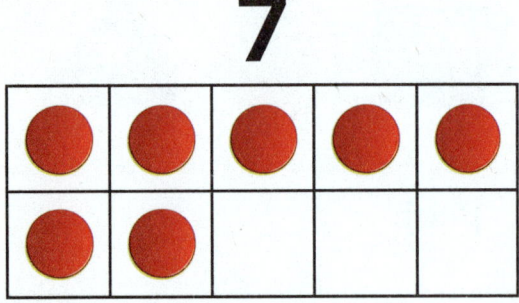

six [seis]

6

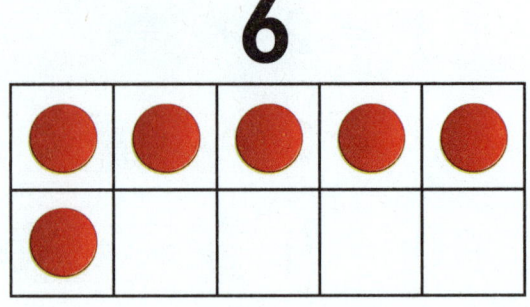

seventeen [diecisiete]

17

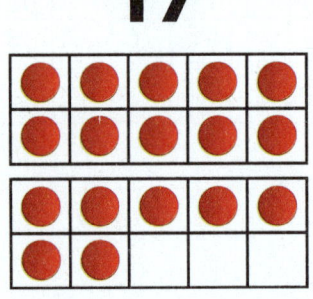

sixteen [dieciséis]

16

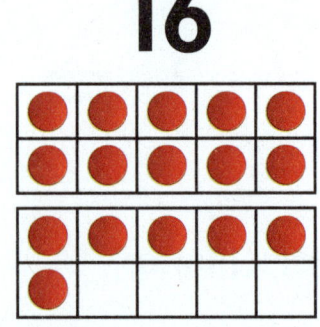

shorter [corta]

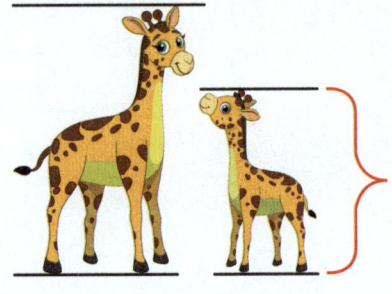

slide [deslizar]

sort [ordenar]

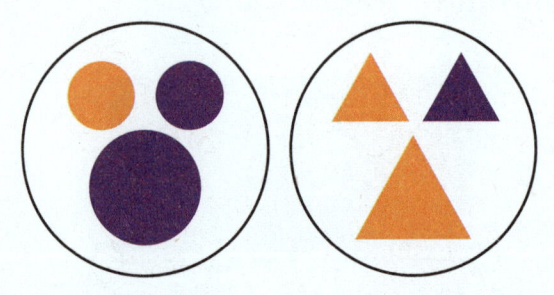

side [lado]

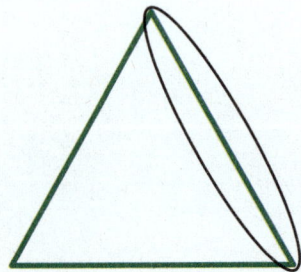

A10

sphere [esfera]

square [cuadrado]

stack [apilar]

subtract [restar]

$$3 - 1 = 2$$

subtraction sentence
[oración de resta]

$$4 - 1 = 3$$

take apart [desmontar]

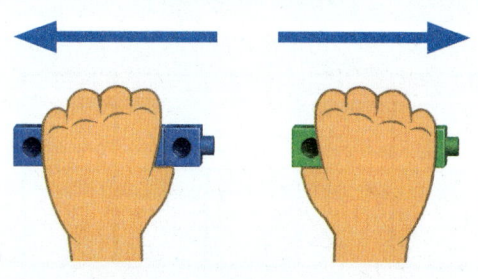

take away [quitar]

taller [más alto]

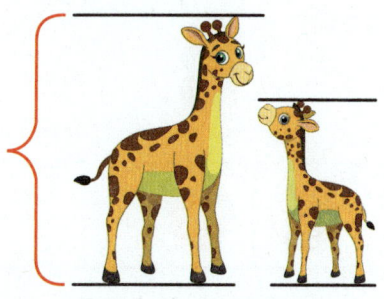

ten [diez]

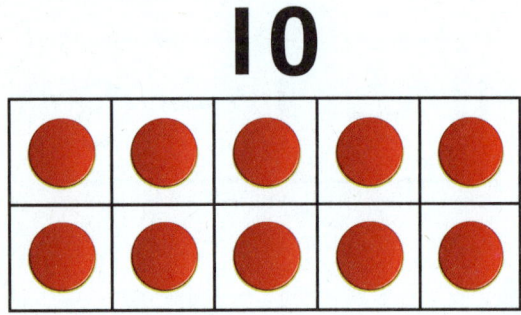

10

ten frame [diez marco]

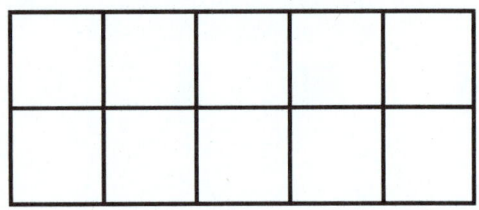

thirteen [trece]

13

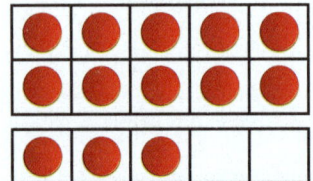

three [tres]

3

three-dimensional shape
[forma tridimensional]

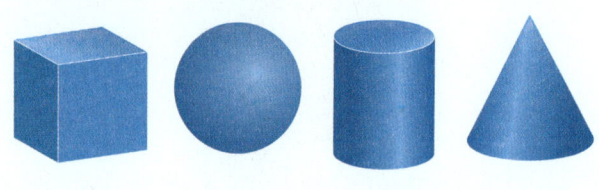

triangle [triángulo]

twelve [doce]

12

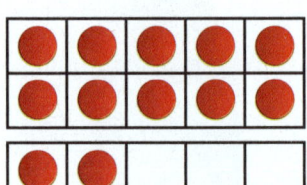

twenty [veinte]

20

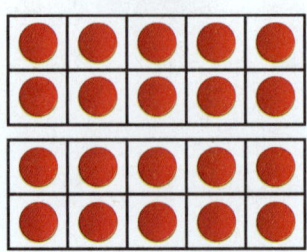

two [dos]

two-dimensional shape
[forma bidimensional]

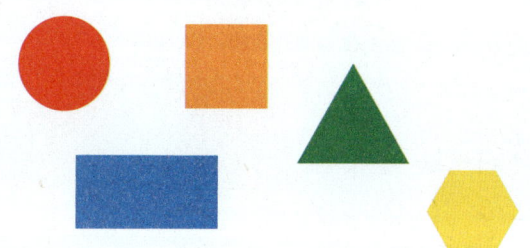

V

vertex [vértice]

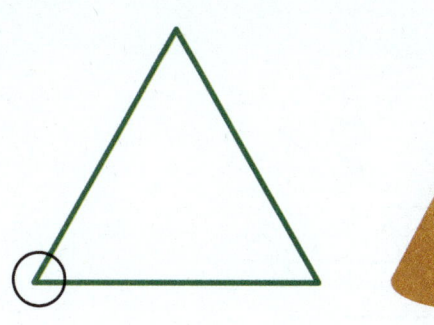

vertices [vértices]

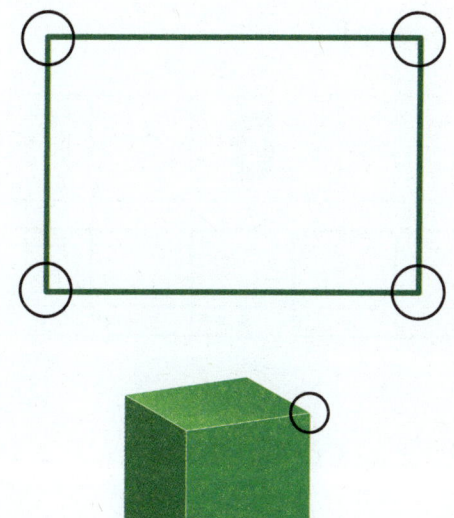

W

weight [peso]

whole [todo]

Z

zero   **[cero]**

0

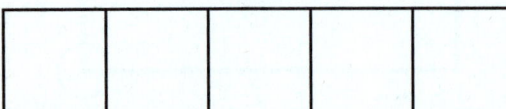

Index

A

Above (position), 628–630, 632
Addition
    add to, 275–280
    using group of 5, 305–310
    partner numbers in
        to 5, 299–304
        finding numbers for, 287–292
        to make 10, 311–316
    practicing, 299–304
    put together, 281–286
    using related facts within 5, 361–366
    understanding, 269–274
Addition patterns, with 0 and 1, 293–298
Addition sentence, 275–280 (*See also* Addition)
***Ants at the Picnic* (story),** 461–466
Apply and Grow: Practice, *In every lesson. For example, see:* 5, 61, 99, 173, 215, 271, 327, 381, 457, 501
***At the Pond* (story),** 33–38
Attributes, measurable
    capacities, 671–676
    describing objects by, 677–682
    height, 641–646
    length, 647–658
    weight, 659–670

B

Balance scale, 665–670
Behind (position), 628–631
Below (position), 628–632
Beside (position), 628–632
***Bugs, Bugs, Bugs* (story),** 127–132
Building
    three-dimensional shapes, 621–626
    two-dimensional shapes, 583–588

C

Capacities
    comparing, 671–676
    describing objects by, 677–682
Castle tower, building
    with three-dimensional shapes, 624
    with two-dimensional shapes, 588
Cat, drawing with two-dimensional shapes, 576
Categories
    classifying objects into, 189–194
    comparing number of objects in, 195–200
Chapter Practice, *In every chapter. For example, see:* 53–56, 91–94, 165–168, 203–206, 263–266, 319–322, 369–372, 447–452, 493–496, 537–540
Chart, hundred
    for counting by ones, 511–516, 523, 524, 527
    for counting by tens
        to 100, 517–522
        within 100, 523, 524, 527
        from number within 100, 529–534
Circles
    building, 583, 585
    drawing cat with, 576
    drawing robot with, 574
    identifying and describing, 571–576, 597
Classifying
    comparing numbers of objects in, 195–200
    objects into categories, 189–194
Column, on hundred chart, 511
Common Errors, *Throughout. For example, see:* T-10, T-140, T-172, T-216, T-332, T-386, T-569

Common Misconceptions, *Throughout. For example, see:* 189, T-192, T-294, T-344, T-671

Comparing capacities, 671–676

Comparing heights, 641–646

Comparing lengths, 647–652

    using numbers, 653–658

Comparing numbers

    to 5 (up to 5), 83–88

    to 10 (up to 10), 183–188

    to 20 (up to 20), 485–490

    equal groups, 59–64

    in groups to 5

        using counting, 77–82

        greater than, 65–70

        less than, 71–76

    in groups to 10

        using counting, 177–182

        using matching, 171–176

    objects in categories, 195–200

Comparing weights, 659–664

    using numbers, 665–670

Composing

    6, 225–230

    7, 231–236

    8, 237–242

    9, 243–248

    10, 249–254

    numbers to 5, 213–218

    numbers to 10, using group of 5, 255–260

Cones

    building, 621–626

    identifying and describing, 597–602, 615–620

    position, based on other objects, 627–632

    roll, stack, or slide sorting of, 603–608

Counting

    1 and 2, 3–8

    3 and 4, 15–20

    5, 27–32

    6, 97–102

    7, 109–114

    8, 121–126

    9, 133–138

    10, 145–150

    11 and 12, 385–390

    13 and 14, 397–402

    15, 409–414

    16 and 17, 421–426

    18 and 19, 433–438

    20, 455–460

    comparing using

        for groups to 5, 77–82

        for groups to 10, 177–182

        for numbers to 20, 485–490

        for objects in categories, 195–200

    finding how many using, 467–472

    forward to 20, from any number, 473–478

    by ones

        to 30, 499–504

        to 50, 505–510

        to 100, 511–516

        within 100, 523–528

    by tens, 523–528

        to 100, 517–522

        from number within 100, 529–534

Counting and ordering

    to 20, 479–484

    numbers to 5, 45–50

    numbers to 10, 157–162

Cross-Curricular Connections, *In every lesson. For example, see:* T-7, T-75, T-131, T-253, T-341, T-443, T-527, T-587, T-613, T-645

Cubes

    building, 621–626

    identifying and describing, 597–602, 609–614

    position, based on other objects, 627–632

    roll, stack, or slide sorting of, 603–608

Cumulative Practice, 207–210, 373–376, 541–544, 689–692

Curved surfaces

    of cones, 616–620

    of cylinders, 616–620

    of spheres, 610–614

Curves
  of circles, 572–576
  of two-dimensional shapes, 547–552
Cylinders
  building, 621–626
  identifying and describing, 597–602, 615–620
  position, based on other objects, 627–632
  roll, stack, or slide sorting of, 603–608

D

Decomposing
  6, 225–230
  7, 231–236
  8, 237–242
  9, 243–248
  10, 249–254
  numbers to 5, 213–218
  numbers to 10, using group of 5, 255–260
  taking apart in subtraction, 337–342
Differentiation, *See* Scaffolding Instruction
Drawing, with two-dimensional shapes
  cat, 576
  house, 568
  robot, 574

E

Eight (8)
  composing and decomposing, 237–242
  modeling and counting, 121–126
  understanding and writing, 127–132
Eighteen (18)
  counting and writing, 433–438
  understanding, 439–444
Eleven (11)
  counting and writing, 385–390
  understanding, 391–396
ELL Support, *In every lesson. For example, see:* T-3, T-59, T-98, T-174, T-261, T-340, T-445, T-517, T-580, T-650
Equal groups, showing and identifying, 59–64

Equal sign (=)
  in add to problems, 276–280
  in put together problems, 282–286
  in take apart problems, 338–342
  in take from problems, 332–336
Equations
  addition, 269–274 (*See also* Addition)
  subtraction, 331–336 (*See also* Subtraction)
Explore and Grow, *In every lesson. For example, see:* 3, 59, 97, 171, 213, 269, 325, 379, 455, 499

F

Fewer
  in groups to 5, 65–70, 77–82
  in groups to 10, 177–182
  in numbers to 5, 83–88
  in numbers to 10, 183–188
  in numbers to 20, 485–490
Fifteen (15)
  counting and writing, 409–414
  understanding, 415–420
Fifty (50), counting by ones to, 505–510
Five (5)
  comparing groups of objects (up to 5)
    using counting, 77–82
    greater than, 65–70
    less than, 71–76
  comparing numbers (up to 5), 83–88
  counting and ordering numbers to, 45–50
  group of
    addition using, 305–310
    composing and decomposing using, 255–260
    subtraction using, 355–360
  modeling and counting, 27–32
  number bonds to represent numbers to, 219–224
  partner numbers to, 213–218, 299–304
  related facts within, adding or subtracting using, 361–366

subtraction within, 349–354
understanding and writing, 33–38

Five frame
    for adding partner numbers to 5,
        299–300, 303
    for addition patterns with 0 and 1, 293
    for comparing number of objects, 65, 71,
        77, 78
    for concept of zero, 39, 40
    for counting and ordering numbers, 46,
        49
    for modeling and counting
        3 and 4, 15–20
        5, 27–32
        11 and 12, 385, 392–396
        13 and 14, 397, 404, 405, 407, 408
        15, 409
    for showing partner numbers, 213
    for subtraction patterns, 343, 344, 347
    for subtraction within 5, 350, 354
Flat shapes, 598–602
Flat surfaces
    of cones, 616–620
    of cubes, 610–614
    of cylinders, 616–620
Formative Assessment, *Throughout. For*
    example, see: T-6, T-130, T-174,
    T-228, T-284, T-350, T-386, T-470,
    T-568, T-624
Four (4)
    modeling and counting, 15–20
    understanding and writing, 21–26
Fourteen (14)
    counting and writing, 397–402
    understanding, 403–408

G

Games, *In every chapter. For example, see:* 52,
    90, 164, 202, 262, 318, 368, 446,
    492, 536
Greater than, 65–70
    in groups to 5, 65–70, 77–82

in groups to 10, 177–182
in numbers to 5, 83–88
in numbers to 10, 183–188
in numbers to 20, 485–490
Groups
    of 5
        addition using, 305–310
        composing and decomposing using,
           255–260
        subtraction using, 355–360
    of 10, identifying, 379–384
    addition in, 269–274
        add to, 275–280
        put together, 281–286
    comparing numbers in
        to 5, 83–88
        using counting, for groups to 5, 77–82
        using counting, for groups to 10,
           177–182
        equal, 59–64
        greater than, 65–70
        less than, 71–76
        using matching, in groups to 10,
           171–176
    counting how many objects in, 467–472
    equal, comparing and identifying, 59–64
    subtraction in
        take apart, 337–342
        take away, 325–330
        take from, 331–336

H

Heavier, 659–664
Heights
    comparing, 641–646
    describing objects by, 677–682
Hexagons
    building, 583, 585
    building house with, 586
    drawing cat with, 576
    drawing robot with, 574
    identifying and describing, 571–576, 597
    joining shapes to make, 578, 581
    joining to make shapes, 579–582

Holds less

    comparing capacities, 671–676

    describing objects by, 677–682

Holds more

    comparing capacities, 671–676

    describing objects by, 677–682

House

    building with two-dimensional shapes, 586

    drawing with two-dimensional shapes, 568

How many, finding

    addition for, 269–275

    counting for, 467–472

    subtraction for (how many left), 325–330

Hundred chart

    for counting by ones, 511–516, 523, 524, 527

    for counting by tens

        to 100, 517–522

        within 100, 523, 524, 527

        from number within 100, 529–534

I

In all, addition for finding how many, 269–274

In front of (position), 627–630, 632

In the Water **(story),** 151–156

J

Joining

    in addition

        add to, 275–280

        put together, 281–286

    two-dimensional shapes, 577–582

L

Learning Target, *In every lesson. For example, see:* 3, 59, 97, 171, 213, 269, 325, 379, 455, 499

Left, how many, 325–330 (*See also* Subtraction)

Lengths

    comparing, 647–652

        using numbers, 653–658

    describing objects by, 677–682

Less, capacity (holds less)

    comparing, 671–676

    describing objects by, 677–682

Less than

    in groups to 5, 71–82

    in groups to 10, 177–182

    in numbers to 5, 83–88

    in numbers to 10, 183–188

    in numbers to 20, 485–490

Lighter, 659–664

Linking cubes

    for adding

        partner numbers, 287–289, 291

        partner numbers to make 10, 311, 313, 316

        put together, 282, 283, 285

        using related facts, 361

    for comparing lengths, 653–658

    for comparing numbers to 20, 485

    for comparing weights, 666–670

    for counting

        1 and 2, 11, 13

        3 and 4, 23, 25

        5, 35, 37

        6, 105, 107

        7, 117, 119

        8, 129, 131

        9, 141, 143

        10, 153, 155

        11 and 12, 385, 387, 389

        13 and 14, 397, 399, 401

        15, 409, 411, 413

        16 and 17, 421, 423, 425

        18 and 19, 433, 435, 437

        20, 455

    for counting and ordering

        numbers to 5, 45

        numbers to 20, 479–484

for counting by tens
    within 100, 524, 525, 527
    from number within 100, 529
for counting forward to 20, 473
for counting to 100 by tens, 518, 519, 521
for finding how many, 467
for groups of 10, 379, 380, 383
for subtracting
    using related facts, 361
    take apart, 338, 339, 341, 342
Longer, 646–652
L-shaped vertices
    of rectangles, 560–564
    of squares, 566–570

M

"Make a 10" strategy, finding partner numbers in, 311–316
Mark, for classifying objects, 190–194
Matching, comparing groups to 10 by, 171–176
Math Musicals, *In every chapter of the Teaching Edition. For example, see:* 146, 184, 214, 338, 456, 548, 616, 654
Measurable attributes
    capacities, 671–676
    describing objects by, 677–682
    height, 641–646
    length, 647–658
    weight, 659–670
Minus sign (−)
    in take apart problems, 338–342
    in take from problems, 331–336
Modeling
    1 and 2, 3–8
    3 and 4, 15–20
    5, 27–32
    6, 97–102
    7, 109–114
    8, 121–126
    9, 133–138

    10, 145–150
    11 and 12, 385–390
    13 and 14, 397–402
    15, 409–414
    16 and 17, 421–426
    18 and 19, 433–438
    20, 455–460
More
    capacity (holds more)
        comparing, 671–676
        describing objects by, 677–682
    in groups to 5, 65–70, 77–82
    in groups to 10, 177–182
    in numbers to 5, 83–88
    in numbers to 10, 183–188
    in numbers to 20, 485–490
Multiple Representations, *Throughout. For example, see:* 22, 256, 270, 276, 300, 312, 326, 332, 338, 428
Music Class (story), 103–108
My Baseball Game (story), 139–144
My Pets (story), 9–14

N

Next to (position), 627, 628, 630, 631
Nine (9)
    composing and decomposing, 243–248
    modeling and counting, 133–138
    understanding and writing, 139–144
Nineteen (19)
    counting and writing, 433–438
    understanding, 439–444
Number(s)
    comparing (*See* Comparing numbers)
    comparing lengths using, 653–658
    comparing weights using, 665–670
    counting and ordering
        to 5, 45–50
        to 10, 157–162
        to 20, 479–484
    modeling and counting
        1 and 2, 3–8
        3 and 4, 15–20

5, 27–32
6, 97–102
7, 109–114
8, 121–126
9, 133–138
10, 145–150
11 and 12, 385–390
13 and 14, 397–402
15, 409–414
16 and 17, 421–426
18 and 19, 433–438
20, 455–460
understanding and writing, 103–108
0, 40, 43
1 and 2, 9–14
3 and 4, 21–26
5, 33–38
6, 103–108
7, 115–120
8, 127–132
9, 139–144
10, 151–156
11 and 12, 385–396
13 and 14, 397–408
15, 409–420
16 and 17, 421–432
18 and 19, 433–444
20, 461–466

Number bonds
for 6, 226–230
for 7, 232–236
for 8, 238–242
for 9, 244–248
for 10, 250–254
for 11, 391
for 13, 403
for 15, 415
for 16, 427
for 18, 439
in addition, 270, 271, 273
for numbers to 5, 219–224
for numbers to 10, using group of 5, 256–260
in subtraction, 338, 339, 341

Number patterns
addition, 293–298
subtraction, 343–348

O

Objects
classifying, 189–194
in groups, addition of, 269–274
add to, 275–280
put together, 281–286
in groups, comparing numbers of
to 5, 83–88
using counting, in groups to 5, 77–82
using counting, in groups to 10, 177–182
equal groups, 59–64
greater than, 65–70
less than, 71–76
using matching, in groups to 10, 171–176
in groups, counting to find how many, 467–472
in groups, subtraction of
take apart, 337–342
take away, 325–330
take from, 331–336
measurable attributes of
capacities, 671–676
describing, 677–682
height, 641–646
length, 647–658
weight, 659–670

One hundred (100)
counting by ones to, 511–516
counting by tens and ones within, 523–528
counting by tens to, 517–522

Ones (1)
in 11 and 12, 391–396
in 13 and 14, 403–408
in 15, 415–420
in 16 and 17, 427–432

Index

in 18 and 19, 439–444

addition patterns with, 293–298

counting by

to 30, 499–504

to 50, 505–510

to 100, 511–516

within 100, 523–528

groups of ten and, 380–384

modeling and counting, 3–8

understanding and writing, 9–14

Ordering numbers

to 5, 45–50

to 10, 157–162

to 20, 479–484

P

Part(s)

in composing and decomposing

6, 225–230

7, 231–236

8, 237–242

9, 243–248

10, 249–254

numbers to 5, 213–218

numbers to 10, using group of 5, 255–260

number bonds and, 219–224

Partner numbers

adding

finding numbers for, 287–292

to make 10, 311–316

numbers to 5, 299–304

for composing and decomposing

6, 225–230

7, 231–236

8, 237–242

9, 243–248

10, 249–254

numbers to 5, 213–218

Pattern blocks, for making shapes, 578–582

Patterns

addition, with 0 and 1, 293–298

subtraction, 343–348

Performance Task, *In every chapter. For example, see:* 51, 89, 163, 201, 261, 317, 367, 445, 491, 535

Plus sign (+)

in add to problems, 275–280

in put together problems, 281–286

Positions, of solid shapes, 627–632

Practice, *In every lesson. For example, see:* 7–8, 63–64, 101–102, 175–176, 217–218, 273–274, 329–330, 383–384, 459–460, 503–504

Problem Types, *Throughout. For example, see:*

add to, result unknown, 278, 302, 319, 374, 378, 562, 683

put together,

both addends unknown, 290, 302, 304, 317, 320, 366, 372

total unknown, 282, 286, 302, 317, 320, 406, 445, 498, 568, 593

take apart,

both addends unknown, 337, 340, 354, 364, 367, 370, 372, 376

total unknown, 352

take from, result unknown, 332, 346, 348, 358, 360, 367

Put together

addition using, 281–286

composing using

6, 225–230

7, 231–236

8, 237–242

9, 243–248

10, 249–254

numbers to 5, 213–218

numbers to 10, using group of 5, 255–260

R

Rainy Day **(story),** 115–120

Reading, *Throughout. For example, see:* T-7, T-149, T-181, T-235, T-279, T-359, T-431, T-515, T-607, T-663

Rectangles
>building, 583, 584, 587
>building house with, 586
>drawing cat with, 576
>drawing house with, 568
>drawing robot with, 574
>identifying and describing, 559–564, 597
>joining squares to make, 577, 581
>joining triangles to make, 579

Related facts, adding or subtracting using, 361–366

Response to Intervention, *Throughout. For example, see:* T-1B, T-61, T-211B, T-289, T-345, T-377B, T-457, T-497B, T-611, T-679

Robot, drawing with two-dimensional shapes, 574

Rocket shapes, creating with pattern blocks, 580

Roll/rolling, by solid shapes, 603–608

Row, on hundred chart, 511

S

Same as (equal groups), 59–64

Scaffolding Instruction, *In every lesson. For example, see:* T-5, T-79, T-111, T-233, T-283, T-363, T-405, T-513, T-549, T-649

Scale, for comparing weights, 665–670

Seven (7)
>composing and decomposing, 231–236
>modeling and counting, 109–114
>understanding and writing, 115–120

Seventeen (17)
>counting and writing, 421–426
>understanding, 427–432

Shapes, *See also* specific shapes
>three-dimensional
>>building, 621–626
>>describing, 597–608
>>identifying, 597–602
>>positions of, 627–632
>>roll, stack, or slide sorting of, 603–608

>two-dimensional
>>building, 583–588
>>curves of, 547–552
>>describing, 547–552, 597–602
>>drawing cat with, 576
>>drawing house with, 568
>>drawing robot with, 574
>>identifying, 597–602
>>joining, 577–582
>>sides of, 548–552
>>vertices of, 548–552

Shorter
>height, 641–646
>length, 646–652

Show how you know, 84, 94, 183, 205, 302, 654, 686

Sides
>in building shapes, 583–588
>of hexagons, 572–576
>of rectangles, 560–564
>of squares, 566–570
>of triangles, 554–558, 584
>of two-dimensional shapes, 548–552

Six (6)
>composing and decomposing, 225–230
>modeling and counting, 97–102
>understanding and writing, 103–108

Sixteen (16)
>counting and writing, 421–426
>understanding, 427–432

Slide/sliding, by solid shapes, 603–608

Solid shapes, 598–602
>building, 621–626
>positions of, 627–632
>roll, stack, or slide sorting of, 603–608

Sorting
>measurable attributes, 677–682
>rectangles, 559–564
>squares, 565–570
>three-dimensional shapes, 603–608
>triangles, 553–558
>two-dimensional shapes, 547–552

Spheres

    building, 621–626

    identifying and describing, 597–602, 609–614

    position, based on other objects, 627–632

    roll, stack, or slide sorting of, 603–614

Squares

    building, 583, 587

    building house with, 586

    drawing house with, 568

    identifying and describing, 565–570, 597

    joining to make shapes, 577, 579–582

Stacks, of solid shapes, 603–608

Stories

    Ants at the Picnic, 461–466

    At the Pond, 33–38

    Bugs, Bugs, Bugs, 127–132

    Music Class, 103–108

    My Baseball Game, 139–144

    My Pets, 9–14

    Rainy Day, 115–120

    In the Water, 151–156

    We Go Camping, 21–26

Straight sides

    of hexagons, 572–576

    of rectangles, 560–564

    of squares, 566–570

    of triangles, 554–558, 584

    of two-dimensional shapes, 548–552

Subtraction

    using group of 5, 355–360

    practicing, within 5, 349–354

    using related facts within 5, 361–366

    take apart, 337–342

    take away, 325–330

    take from, 331–336

    understanding, 325–330

Subtraction patterns, 343–348

Subtraction sentence, 331–336 (*See also* Subtraction)

Success Criteria, *In every lesson. For example, see:* T-3, T-65, T-189, T-237, T-299, T-361, T-421, T-511, T-583, T-647

Symbols

    equal sign ($=$)

        in add to problems, 276–280

        in put together problems, 282–286

        in take apart problems, 338–342

        in take from problems, 332–336

    minus sign ($-$)

        in take apart problems, 338–342

        in take from problems, 331–336

    plus sign ($+$)

        in add to problems, 275–280

        in put together problems, 281–286

T

Take apart (decomposing)

    6, 225–230

    7, 231–236

    8, 237–242

    9, 243–248

    10, 249–254

    numbers to 5, 213–218

    numbers to 10, using group of 5, 255–260

    subtraction using, 337–342

Take away, 325–330

Take from, 331–336

Taller, 641–646

Ten frames

    for adding

        using group of 5, 305–310

        partner numbers making 10, 312, 313, 315, 316

    for comparing numbers up to 20, 485, 486, 489

    for composing and decomposing, 256, 257

    for counting and ordering numbers

        to 5, 157–159

        to 20, 479–484

    for counting forward to 20, 473

    for modeling and counting

        6, 97–102

        7, 109–114

8, 121–126

9, 133–138

10, 145–150

11 and 12, 385, 392–396

13 and 14, 397, 404–408

15, 409, 416–420

16 and 17, 421, 428–432

18 and 19, 433, 440–444

20, 455–460

for subtracting, using group of 5, 355–357, 359, 360

for subtraction patterns, 344–345

Tens (10)

in 11 and 12, 391–396

in 13 and 14, 403–408

in 15, 415–420

in 16 and 17, 427–432

in 18 and 19, 439–444

adding partner numbers to make, 311–316

comparing groups of objects (up to 10)

by counting, 177–182

by matching, 171–176

comparing numbers (up to 10), 183–188

composing and decomposing, 249–254

composing and decomposing numbers to, using group of 5, 255–260

counting and ordering numbers to, 157–162

counting by

to 100, 517–522

within 100, 523–528

from number within 100, 529–534

groups of, identifying, 379–384

modeling and counting, 145–150

understanding and writing, 151–156

Think and Grow, *In every lesson. For example, see:* 4, 60, 98, 172, 214, 270, 326, 380, 456, 500

Think and Grow: Modeling Real Life, *In every lesson. For example, see:* 6, 62, 100, 174, 216, 272, 328, 382, 458, 502

Thirteen (13)

counting and writing, 397–402

understanding, 403–408

Thirty (30), counting by ones to, 499–504

Three (3)

modeling and counting, 15–20

understanding and writing, 21–26

Three-dimensional shapes

building, 621–626

describing, 597–608

identifying, 597–602

positions of, 627–632

roll, stack, or slide sorting of, 603–608

Totem pole, building, 626

Triangles

building, 583, 584, 586–588

building house with, 586

drawing cat with, 576

drawing robot with, 574

identifying and describing, 553–558, 597

joining to make shapes, 578–582

Twelve (12)

counting and writing, 385–390

understanding, 391–396

Twenty (20)

comparing numbers (up to 20), 485–490

counting, 455–466

counting forward from any number to, 473–478

modeling, 455–460

ordering numbers to, 479–484

writing, 461–466

Two (2)

modeling and counting, 3–8

understanding and writing, 9–14

Two-dimensional shapes

building, 583–588

curves of, 547–552

describing, 547–552, 597–602

drawing cat with, 576

drawing house with, 568

drawing robot with, 574

Index

identifying, 597–602
joining, 577–582
sides of, 548–552
vertices of, 548–552

Vertex (vertices)
  in building shapes, 583–588
  of hexagons, 572–576
  of rectangles, 560–564
  of squares, 566–570
  of triangles, 554–558, 584
  of two-dimensional shapes, 548–552

We Go Camping (story), 21–26
Weights
  comparing, 659–664
    using numbers, 665–670
  describing objects by, 677–682
Whole
  composing and decomposing
    6, 225–230
    7, 231–236
    8, 237–242
    9, 243–248
    10, 249–254
    numbers to 5, 213–218
    numbers to 10, using group of 5,
      255–260
  number bonds and, 219–224

Writing numbers
  0, 40, 43
  1 and 2, 9–14
  3 and 4, 21–26
  5, 33–38
  6, 103–108
  7, 115–120
  8, 127–132
  9, 139–144
  10, 151–156
  11 and 12, 385–390
  13 and 14, 397–402
  15, 409–414
  16 and 17, 421–426
  18 and 19, 433–438
  20, 461–466

Zero (0)
  addition patterns with, 293–298
  concept of, 39–44
  writing, 40, 43

Credits